T0327585

SMART GRID

A complete list of titles in the IEEE Press Series on Power Engineering appears
at the end of this book.

SMART GRID
Fundamentals of Design and Analysis

James Momoh

IEEE
PRESS
SERIES
ON POWER
ENGINEERING

Mohamed E. El-Hawary, *Series Editor*

IEEE PRESS

A JOHN WILEY & SONS, INC., PUBLICATION

Published by John Wiley & Sons, Inc., Hoboken, New Jersey. All rights reserved.
Published simultaneously in Canada.

For general information on our other products and services or for technical support, please contact our Customer Care Department within the United States at (800) 762-2974, outside the United States at (317) 572-3993 or fax (317) 572-4002.

Wiley also publishes its books in a variety of electronic formats. Some content that appears in print may not be available in electronic formats. For more information about Wiley products, visit our web site at www.wiley.com.

Library of Congress Cataloging-in-Publication Data:

Momoh, James A., 1950-
  Smart grid : fundamentals of design and analysis / James Momoh.
      p. cm.
  ISBN 978-0-470-88939-8 (hardback)
  1. Electric power distribution–Automation.   I. Title.
  TK3226.M588 2012
  333.793'2–dc23

                                        2011024774

CONTENTS

8 INTEROPERABILITY, STANDARDS, AND CYBER SECURITY   160

9 RESEARCH, EDUCATION, AND TRAINING FOR THE SMART GRID   176

PREFACE

The term "smart grid" defines a self-healing network equipped with dynamic optimization techniques that use real-time measurements to minimize network losses, maintain voltage levels, increase reliability, and improve asset management. The operational data collected by the smart grid and its sub-systems will allow system operators to rapidly identify the best strategy to secure against attacks, vulnerability, and so on, caused by various contingencies. However, the smart grid first depends upon identifying and researching key performance measures, designing and testing appropriate tools, and developing the proper education curriculum to equip current and future personnel with the knowledge and skills for deployment of this highly advanced system.

The objective of this book is to equip readers with a working knowledge of fundamentals, design tools, and current research, and the critical issues in the development and deployment of the smart grid. The material presented in its eleven chapters is an outgrowth of numerous lectures, conferences, research efforts, and academic and industry debate on how to modernize the grid both in the United States and worldwide. For example, Chapter 3 discusses the optimization tools suited to managing the randomness, adaptive nature, and predictive concerns of an electric grid. The general purpose Optimal Power Flow, which takes advantage of a learning algorithm and is capable of solving the optimization scheme needed for the generation, transmission, distribution, demand response, reconfiguration, and the automation functions based on real-time measurements, is explained in detail.

I am grateful to several people for their help during the course of writing this book. I acknowledge Keisha D'Arnaud, a dedicated recent graduate student at the Center for Energy Systems and Control, for her perseverance and support in the several iterations needed to design the text for a general audience. I thank David Owens, Senior Executive Vice President of the Edison Electric Institute, and Dr. Paul Werbos, Program Director of the Electrical, Communication and Cyber Systems (ECCS), National Science Foundation (NSF), for encouraging and supporting my interest in unifying my knowledge of systems through computational intelligence to address complex power system problems where traditional techniques have failed. Their support was especially valuable during my stint at NSF as a Program Director in ECCS from 2001 to 2004. I am also grateful for the Small Grant Expository Research award granted by the NSF to develop the first generation of Dynamic Stochastic Optimal Power flow, a general purpose tool for use in smart grid design and development.

I thank my family for their encouragement and support. I am grateful to my students and colleagues at the Center for Energy Systems and Control, who, as audience and enthusiasts, let me test and refine my ideas in the smart grid, and also for honorary invited presentations to top utility executive management in addressing the emergence of the smart grid across the country. All these exposures rekindled my interest in the design and development of the grid for the future.

JAMES MOMOH

1

SMART GRID ARCHITECTURAL DESIGNS

1.1 INTRODUCTION

Today's electric grid was designed to operate as a vertical structure consisting of generation, transmission, and distribution and supported with controls and devices to maintain
reliability, stability, and efficiency. However, system operators are now facing new challenges including the penetration of RER in the legacy system, rapid technological change,
and different types of market players and end users. The next iteration, the smart grid,
will be equipped with communication support schemes and real-time measurement techniques to enhance resiliency and forecasting as well as to protect against internal and
external threats. The design framework of the smart grid is based upon unbundling and
restructuring the power sector and optimizing its assets. The new grid will be capable of:

- Handling uncertainties in schedules and power transfers across regions
- Accommodating renewables
- Optimizing the transfer capability of the transmission and distribution networks
  and meeting the demand for increased quality and reliable supply
- Managing and resolving unpredictable events and uncertainties in operations and
  planning more aggressively.

Smart Grid: Fundamentals of Design and Analysis, First Edition. James Momoh.
© 2012 Institute of Electrical and Electronics Engineers. Published 2012 by John Wiley & Sons, Inc.

TABLE 1.1. Comparison of Today's Grid vs. Smart Grid [4]

Preferred Characteristics	Today's Grid	Smart Grid
Active Consumer Participation	Consumers are uninformed and do not participate	Informed, involved consumers—demand response and distributed energy resources
Accommodation of all generation and storage options	Dominated by central generation—many obstacles exist for distributed energy resources interconnection	Many distributed energy resources with plug-and-play convenience focus on renewables
New products, services, and markets	Limited, poorly integrated wholesale markets; limited opportunities for consumers	Mature, well-integrated wholesale markets; growth of new electricity markets for consumers
Provision of power quality for the digital economy	Focus on outages—slow response to power quality issues	Power quality a priority with a variety of quality/price options—rapid resolution of issues
Optimization of assets and operates efficiently	Little integration of operational data with asset management—business process silos	Greatly expanded data acquisition of grid parameters; focus on prevention, minimizing impact to consumers
Anticipating responses to system disturbances (self-healing)	Responds to prevent further damage; focus on protecting assets following a fault	Automatically detects and responds to problems; focus on prevention, minimizing impact to consumers
Resiliency against cyber attack and natural disasters	Vulnerable to malicious acts of terror and natural disasters; slow response	Resilient to cyber attack and natural disasters; rapid restoration capabilities

1.2   TODAY'S GRID VERSUS THE SMART GRID

As mentioned, several factors contribute to the inability of today's grid to efficiently meet the demand for reliable power supply. Table 1.1 compares the characteristics of today's grid with the preferred characteristics of the smart grid.

1.3   ENERGY INDEPENDENCE AND SECURITY ACT OF 2007: RATIONALE FOR THE SMART GRID

The Energy Independence and Security Act of 2007 (EISA) signed into law by President George W. Bush vividly depicts a smart grid that can predict, adapt, and reconfigure itself efficiently and reliably. The objective of the modernization of the U.S. grid as outlined in the Act is to maintain a reliable and secure electricity [2] infrastructure that

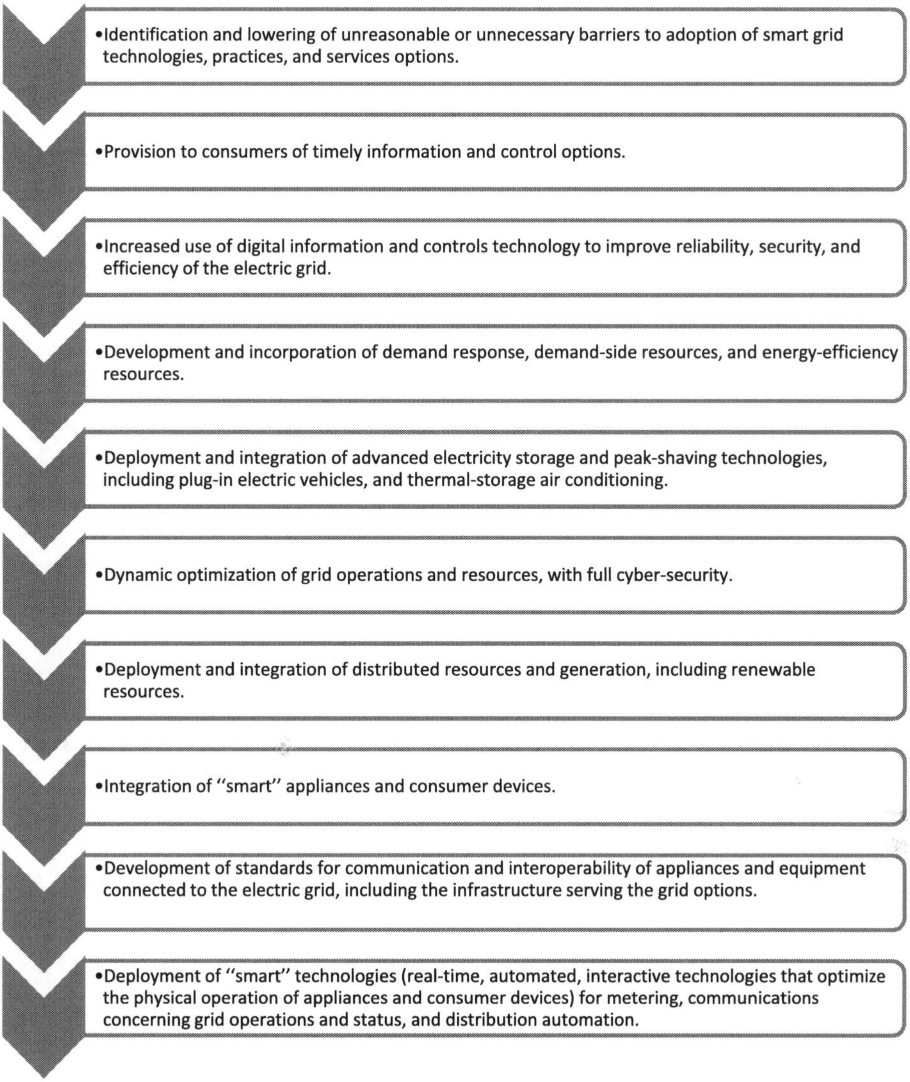

- Identification and lowering of unreasonable or unnecessary barriers to adoption of smart grid technologies, practices, and services options.

- Provision to consumers of timely information and control options.

- Increased use of digital information and controls technology to improve reliability, security, and efficiency of the electric grid.

- Development and incorporation of demand response, demand-side resources, and energy-efficiency resources.

- Deployment and integration of advanced electricity storage and peak-shaving technologies, including plug-in electric vehicles, and thermal-storage air conditioning.

- Dynamic optimization of grid operations and resources, with full cyber-security.

- Deployment and integration of distributed resources and generation, including renewable resources.

- Integration of "smart" appliances and consumer devices.

- Development of standards for communication and interoperability of appliances and equipment connected to the electric grid, including the infrastructure serving the grid options.

- Deployment of "smart" technologies (real-time, automated, interactive technologies that optimize the physical operation of appliances and consumer devices) for metering, communications concerning grid operations and status, and distribution automation.

Figure 1.1. Rationale for the smart grid.

will meet future demand growth. Figure 1.1 illustrates the features needed to facilitate the development of an energy-efficient, reliable system.

The Act established a Smart Grid Task Force, whose mission is "to insure awareness, coordination and integration of the diverse activities of the DoE Office and elsewhere in the Federal Government related to smart-grid technologies and practices" [1]. The task force's activities include research and development; development of widely accepted standards and protocols; the relationship of smart grid technologies and practices to electric utility regulation; the relationship of smart grid technologies and practices to infrastructure development, system reliability, and security; and the relationship of smart grid technologies and practices to other facets of electricity supply, demand, transmission, distribution, and policy. In response to the legislation, the U.S. research and education community is actively engaged in:

1. Smart grid research and development program
2. Development of widely accepted smart grid standards and protection
3. Development of infrastructure to enable smart grid deployment
4. Certainty of system reliability and security
5. Policy and motivation to encourage smart grid technology support for generation, transmission and distribution

As Figure 1.2 shows, there are five key aspects of smart grid development and deployment.

1.4   COMPUTATIONAL INTELLIGENCE

Computational intelligence is the term used to describe the advanced analytical tools needed to optimize the bulk power network. The toolbox will include heuristic, evolution programming, decision support tools, and adaptive optimization techniques.

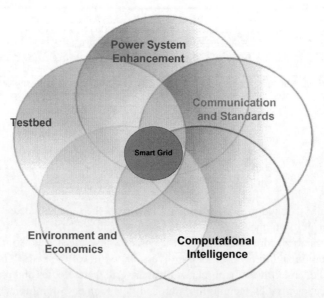

Figure 1.2. Five key aspects of smart grid development.

1.5 POWER SYSTEM ENHANCEMENT

Policy-makers assume that greatly expanded use of renewable energy [4,5] resources in the United States will help to offset the impacts of carbon emissions from thermal and fossil energy, meet demand uncertainty, and to some extent, increase reliability of delivery.

1.6 COMMUNICATION AND STANDARDS

Since planning horizons can be short as an hour ahead, the smart grid's advanced automations will generate vast amounts of operational data in a rapid decision-making environment. New algorithms will help it become adaptive and capable of predicting with foresight. In turn, new rules will be needed for managing, operating, and marketing networks.

1.7 ENVIRONMENT AND ECONOMICS

Based on these desired features, an assessment of the differences in the characteristics of the present power grid and the proposed smart grid is needed to highlight characteristics of the grid and the challenges. When fully developed the smart grid system will allow customer involvement, enhance generation and transmission with tools to allow minimization of system vulnerability, resiliency, reliability, adequacy and power quality. The training tools and capacity development to manage and operate the grids and hence crate new job opportunities is part of the desired goals of the smart grid evolution which will be tested using test-bed. To achieve the rapid deployment of the grids test bed and research centers need to work across disciplines to build the first generation of smart grid.

By focusing on security controls rather than individual vulnerabilities and threats, utility companies and smart-grid technology vendors can remediate the root causes that lead to vulnerabilities. However, security controls are more difficult and sometimes impossible to add to an existing system, and ideally should be integrated from the beginning to minimize implementation issues. The operating effectiveness of the implemented security controls-base will be assessed routinely to protect the smart grid against evolving threats.

1.8 OUTLINE OF THE BOOK

This book is organized into 10 chapters. Following this chapter's introduction, Chapter 2 presents the smart grid concept, fundamentals, working definitions, and system architecture. Chapter 3 describes the tools using load flow concepts, optimal power flows, and contingencies and Chapter 4 describes those using voltage stability, angle stability, and state estimation. Chapter 5 evaluates the computational intelligence approach as a feature of the smart grid. Chapter 6 explains the pathways design of the smart grid using general purpose dynamic stochastic optimization. Chapter 7 reviews renewable supply and the related issues of variability and probability distribution functions, followed by a discussion of storage technologies, capabilities, and configurations. Demand side managemen (DSM) and demand response, climate change, and tax credits are highlighted for the purpose of evaluating the economic and environmental benefit of renewable energy sources. Chapter 8 discusses the importance of developing national standards, followed by a discussion of interoperability such that the new technologies can easily be adapted to the legacy system without violating operational constraints. The chapter also discusses cyber security to protect both RER and communication infrastructure. Chapter 9 explains the significant research and employment training for attaining full performance and economic benefits of the new technology. Chapter 10 discusses case studies on smart grid development and testbeds to aid deployment. The chapter outlines the grand challenges facing researchers and policy-makers before the smart grid can be fully deployed, and calls for investment and multidisciplinary collaboration. Figure 1.3 is a schematic of the chapters.

1.9 GENERAL VIEW OF THE SMART GRID MARKET DRIVERS

To improve efficiency and reliability, several market drivers and new opportunities suggest that the smart grid must:

1. Satisfy the need for increased integration of digital systems for increased efficiency of the power system. In the restructured environment, the deregulated electric utility industry allows a renovation of the market to be based on system constraints and the seasonal and daily fluctuations in demand. Competitive markets increase the shipment of power between regions, which further strains today's aging grid and requires updated, real-time controls.
2. Handle grid congestion, increase customer participation, and reduce uncertainty for investment. This requires the enhancement of the grid's capability to handle demand reliably.
3. Seamlessly integrate renewable energy systems (RES) and distributed generation. The drastic increase in the integration of cost-competitive distributed generation technologies affects the power system.

In addition to system operators and policy-makers, stakeholders are contributing to the development of the smart grid. Their specific contributions and conceptual understanding of the aspects to be undertaken are discussed below.

1.10 STAKEHOLDER ROLES AND FUNCTION

As in the legacy system, critical attention must be paid to the identification of the stakeholders and how they function in the grid's development. Stakeholders range from

Figure 1.3. Schematic of chapters.

utility and energy producers to consumers, policy-makers, technology providers, and researchers. An important part of the realization of the smart grid is the complete buy-in or involvement of all stakeholders.

Policy-makers are the federal and state regulators responsible for ensuring the cohesiveness of policies for modernization efforts and mediating the needs of all parties. The primary benefit of smart grid development to these stakeholders concerns the mitigation of energy prices, reduced dependence on foreign oil, increased efficiency, and reliability of power supply. Figure 1.4 shows the categories of stakeholders.

Other participants in the development of the smart grid include government agencies, manufacturers, and research institutes. The federal Department of Energy's (DOE)

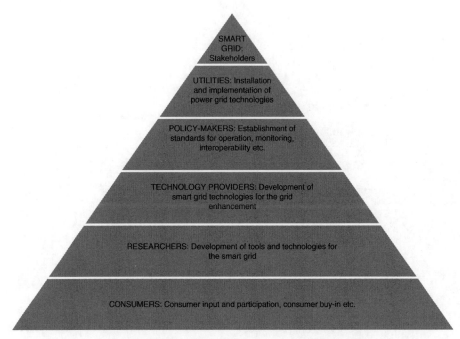

Figure 1.4. Stakeholders and their functions.

National Renewable Energy Laboratory (NREL) and state agencies such as the California Energy Commission and the New York State Energy Research and Development Authority are among the pioneers. In the monograph, "The Smart Grid: An Introduction," the DOE discusses the nature, challenges, opportunities, and necessity for smart grid implementation. It defines the smart grid as technology which "makes this transformation of the electric industry possible by bringing the philosophies, concepts and technologies that enabled the internet to the utility and the electric grid and enables the grid modernization" [1]. The characteristics of the smart grid are two-way digital communication, plug-and-play capabilities, advanced metering infrastructure for integrating customers, facilities for increased customer involvement, interoperability based on standards, and low-cost communication and electronics.

Additional features identified include integration and advancement of grid visualization technology to provide wide-area grid awareness, integrating real-time sensor data, weather information, and grid modeling with geographical information [1].

However, the DOE's definitions in our opinion do not provide measures for addressing uncertainty, predictivity, and foresight. Another federal entity, the Federal Energy Regulatory Commission (FERC), has mandated the development of:

1. **Cyber Security**: require NIST define standard and protocol consistent with the overarching cyber security and reliability requirements of the Energy Independence and Security Act (EISA) and the FERC Reliability Standards.

2. **Intersystem Communications**: Identify standards for common information models for communication among all elements of the bulk power system regional market operators, utilities, demand response aggregators, and customers

3. **Wide-Area Situational Awareness**: Ensure that operators of the nation's bulk power system have the equipment that gives them a complete view of their systems so they can monitor and operate their systems.

4. **Coordination of the bulk power systems with new and emerging technologies:** Identify standards development that help to accommodate the introduction and expansion of renewable resources, demand response, and electricity storage to address several bulk power system challenges. Also identify standards development that help to accommodate another emerging technology, electric transportation.

1.10.1 Utilities

South California Edison (SCE) and other utility companies undertook to reinvent electrical metering. Vendors are migrated to an open standards–based advanced metering infrastructure. These contributions have led to the continual improvement of associated features such as customer service, energy conservation, and economic efficiency.

PEPCO Holdings has been working on an Advanced Metering Infrastructure (AMI). The technology is an integral component of the smart grid [5]. The features proposed include investment in and implementation of innovative, customer-focused technologies and initiatives for efficient energy management, increased pricing options and demand response, reduction of total energy cost and consumption, and reduction of the environmental impacts of electric power consumption.

1.10.2 Government Laboratory Demonstration Activities

Much of the fundamental thinking behind the smart grid concept arose from the DOE's Pacific Northwest National Laboratory (PNNL) more than 20 years ago. In the middle 1980s researchers at PNNL were already designing first-generation data collection systems that were installed in more than 1000 buildings to monitor near real time electricity consumption for every appliance. PNNL developed a broad suite of analytical tools and technologies that resulted in better sensors, improved diagnostics, and enhanced equipment design and operation, from phasor measurement and control at the transmission level to grid-friendly appliances [2]. In January 2006, four years after its first presentation, PNNL unveiled the GridWise Initiative whose objective was the testing of new electric grid technologies [3]. This demonstration project involved 300 homeowners in Washington and Oregon.

The GridWise Alliance manages the GridWise Program in the DOE's Office of Electricity and Energy Assurance. Members include Areva, GE, IBM, Schneider Electric; American Electric Power, Bonneville Power Administration, ConEd, the PJM Interconnection; Battelle, RDS, SAIC, Nexgen, and RockPort Capital Partners [2]. The GridWise Architecture Council [4], a primary advocate for the smart grid, promotes the benefits of improving interoperability between the automation systems needed to enable smart grid applications.

1.10.3   Power Systems Engineering Research Center (PSERC)

The Power Systems Engineering Research Center (PSERC) [6] consists of 13 universities and industrial collaborators involved in research aimed at solving grid problems using state-of-the-art technologies. The direction of PSERC is the development of new strategies, technologies, analytical capabilities, and computational tools for operating and planning practices that will support an adaptive, reliable, and stable power grid.

1.10.4   Research Institutes

The Electric Power Research Institute (EPRI) and university consortium groups have developed software architecture for smart grid development. These tools focus on the development of the grid's technical framework through the integration of electricity systems, communications, and computer controls. The Intelligrid software from EPRI, an open-standard, requirements-based approach for integrating data networks and equipment, enables interoperability between products and systems. It provides methodology, tools, and recommendations for standards and technologies for utility use in planning, specifying, and procuring IT-based systems.

1.10.5   Technology Companies, Vendors, and Manufacturers

IBM is a major player in the provision of information technology (IT) equipment for the smart grid on a global level. In 2008, IBM was chosen to spearhead IT support and services for smart-grid energy-efficiency programs by American Electric Power, Michigan Gas and Electric, and Consumers Energy. IBM serves as the systems integrator for its GridSmart program that displays energy usage and participate in energy-efficiency program. Its Intelligent Power Grid is characterized by increased grid observability with modern data integration and analytics to support advanced grid operation and control, power delivery chain integration, and high-level utility strategic planning functions [7]. Some key characteristics of the Intelligent Power Grid are:

- Grid equipment and assets contain or are monitored by intelligent IP-enabled devices (digital processors).
- Digital communication networks permit the intelligent devices to communicate securely with the utility enterprise and possibly with each other.
- Data from the intelligent devices and many other sources are consolidated to support the transformation of raw data into useful information through advanced analytics.
- Business intelligence and optimization tools provide advanced decision support at both the automatic and human supervisory level.

The data base and architecture consist of five major components: data sources, data transport, data integration, analytics, and optimization. In addition there are means for data distribution which includes publish-and-subscribe middleware, portals, and Web-based services [8].

CISCO has also contributed with its IP architecture. CISCO describes the smart grid as a data communication network integrated with the electrical grid that collects and analyzes data captured in near-real time about power transmission, distribution, and consumption. Predictive information and recommendations to stakeholders are developed based on the data for power management. Integration of the generation, transmission, distribution, and end user components is a critical feature.

There is no one acceptable or universal definition for the smart grid; rather it is function-selected. Below we give a working definition to encompass the key issues of stakeholders and developers.

1.11   WORKING DEFINITION OF THE SMART GRID BASED ON PERFORMANCE MEASURES

A working definition should include the following attributes:

- Assess grid health in real time
- Predict behavior, anticipate
- Adapt to new environments like distributed resources and renewable energy resources
- Handle stochastic demand and respond to smart appliances
- Provide self-correction, reconfiguration, and restoration
- Handle randomness of loads and market participants in real time
- Create more complex interactive behavior with intelligent devices, communication protocols, and standard and smart algorithms to improve smart communication and transportation systems.

In this environment, smart control strategies will handle congestion, instability, or reliability problems. The smart grid will be cyber-secure, resilient, and able to manage shock to ensure durability and reliability. Additional features include facilities for the integration of renewable and distribution resources, and obtaining information to and from renewable resources and plug-in hybrid vehicles. New interface technologies will make data flow patterns and information available to investors and entrepreneurs interested in creating goods and services.

Thus, the working definition becomes:

The smart grid is an advanced digital two-way power flow power system capable of self-healing, and adaptive, resilient, and sustainable, with foresight for prediction under different uncertainties. It is equipped for interoperability with present and future standards of components, devices, and systems that are cyber-secured against malicious attack.

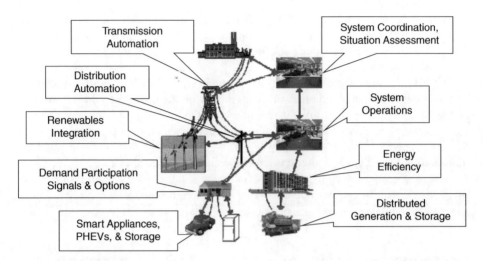

Figure 1.5. DOE representative architecture of the smart grid design (architecture 1).

It is enabled to perform with robust and affordable real-time measurements and enhanced communication technology for data/information transmission. It allows smart appliances and facilitates the deployment of advanced storage technologies including plug-in electric and hybrid vehicles and control options, and supports DSM and demand response schemes.

1.12   REPRESENTATIVE ARCHITECTURE

Several types of architecture have been proposed by the various bodies involved in smart grid development. We present two: one from the DOE and one illustrated by Figure 1.5, which shows how the DOE's proposed smart grid divides into nine areas: transmission automation, system coordination situation assessment, system operations, distribution automation, renewable integration, energy efficiency, distributed generation and storage, demand participation signals and options, and smart appliances, PHEVs, and storage.

Figure 1.6 shows how the second architectural framework is partitioned into subsystems with layers of intelligence and technology and new tools and innovations. It involves bulk power generation, transmission, distribution, and end user level of the electric power system. The function of each component is explained in the next section.

1.13   FUNCTIONS OF SMART GRID COMPONENTS

For the generation level of the power system, smart enhancements will extend from the technologies used to improve the stability and reliability of the generation to intelligent controls and the generation mix consisting of renewable resources.

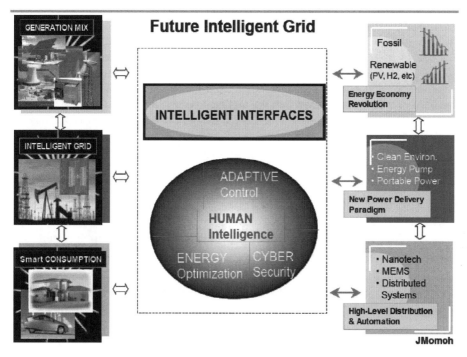

Figure 1.6. The intelligent grid (architecture 2).

1.13.1 Smart Devices Interface Component

Smart devices for monitoring and control form part of the generation components' real time information processes. These resources need to be seamlessly integrated in the operation of both centrally distributed and district energy systems.

1.13.2 Storage Component

Due to the variability of renewable energy and the disjoint between peak availability and peak consumption, it is important to find ways to store the generated energy for later use. Options for energy storage technologies include pumped hydro, advance batteries, flow batteries, compressed air, super-conducting magnetic energy storage, super-capacitors, and flywheels. Associated market mechanisms for handling renewable energy resources, distributed generation, environmental impact and pollution are other components necessary at the generation level.

Associated market mechanism for handling renewable energy resources, distributed generation, environmental impact and pollution has to be introduced in the design of smart grid component at the generation level.

1.13.3 Transmission Subsystem Component

The transmission system that interconnects all major substation and load centers is the backbone of an integrated power system. Efficiency and reliability at an affordable cost continue to be the ultimate aims of transmission planners and operators. Transmission lines must tolerate dynamic changes in load and contingency without service disruptions. To ensure performance, reliability and quality of supply standards are preferred following contingency. Strategies to achieve smart grid performance at the transmission level include the design of analytical tools and advanced technology with intelligence for performance analysis such as dynamic optimal power flow, robust state estimation, real-time stability assessment, and reliability and market simulation tools. Real-time monitoring based on PMU, state estimators sensors, and communication technologies are the transmission subsystem's intelligent enabling tools for developing smart transmission functionality.

1.13.4 Monitoring and Control Technology Component

Intelligent transmission systems/assets include a smart intelligent network, self-monitoring and self-healing, and the adaptability and predictability of generation and demand robust enough to handle congestion, instability, and reliability issues. This new resilient grid has to withstand shock (durability and reliability), and be reliable to provide real-time changes in its use.

1.13.5 Intelligent Grid Distribution Subsystem Component

The distribution system is the final stage in the transmission of power to end users. Primary feeders at this voltage level supply small industrial customers and secondary distribution feeders supply residential and commercial customers. At the distribution level, intelligent support schemes will have monitoring capabilities for automation using smart meters, communication links between consumers and utility control, energy management components, and AMI. The automation function will be equipped with self-learning capability, including modules for fault detection, voltage optimization and load transfer, automatic billing, restoration and feeder reconfiguration, and real-time pricing.

1.13.6 Demand Side Management Component

Demand side management options and energy efficiency options developed for effective means of modifying the consumer demand to cut operating expenses from expensive generators and defer capacity addition.

DSM options provide reduced emissions in fuel production, lower costs, and contribute to reliability of generation. These options have an overall impact on the utility load curve. A standard protocol for customer delivery with two-way information highway technologies as the enabler is needed. Plug-and-play, smart energy buildings and smart homes, demand-side meters, clean air requirements, and customer interfaces for better energy efficiency will be in place.

1.14 SUMMARY

This chapter has discussed the progress made by different stakeholders in the design and development of the smart grid. A working definition of the smart grid was given. Two design architectures and the specific aspects of prospective smart grid function were provided. The next chapters discuss the tools and techniques needed for smart grid analysis and development.

REFERENCES

[1] "The Smart Grid: An Introduction and Smart Grid System Report." Litos Strategic Communication, U.S. Department of Energy, 2009.

[2] L.D. Kinter-Meyer, M.C. Chassin, D.P. Kannberg, et al. "GridWiseTM: The Benefits of a Transformed Energy System." Pacific Northwest National Laboratory, PNNL-14396, 2003.

[3] "Overview of the Smart Grid: Policies, Initiatives and Needs." ISO New England, 2009.

[4] "The Modern Grid Initiative." GridWise Architecture Council, Pacific Northwest National Laboratory, 2008.

[5] "Our Blueprint for the Future." PEPCO Holdings, 2009.

[6] "PSERC Overview, 2008." PSERC, 2008.

[7] J. Taft. "The Intelligent Power Grid." *IBM Global Services*, 2006.

[8] "A National Vision for Electricity's Second 100 Years." Office of Electric Transmission and Distribution, U.S. Department of Energy, 2003.

SUGGESTED READINGS

American Recovery and Reinvestment Act of 2009. Public Law No. 111-5, 2009.

P. Van Doren and J. Taylor. "Rethinking Electricity Restructuring." *Policy Analysis* 2004, 530, 1–8.

EPRI Intelligrid. Electric Power Research Institute, 2001–2010.

"Smart Grid System Report." U.S. Department of Energy, 2009.

The Energy Independence and Security Act of 2007. S. 1419, 90d Congress, 2007.

"The Smart Grid: An Introduction and Smart Grid System Report." U.S. Department of Energy, 2009.

2

SMART GRID COMMUNICATIONS AND MEASUREMENT TECHNOLOGY

2.1   COMMUNICATION AND MEASUREMENT

Because much of the existing transmission and distribution systems in the United States still uses older digital communication and control technology, advanced communication systems for distribution automation, such as Remote Terminal Unit (RTU) [3] and SCADA, are under development as well as innovative tools and software that will communicate with appliances in the home [1]. Ultimately, high-speed, fully integrated, two-way communication technologies will allow the smart grid to be a dynamic, interactive mega-infrastructure for real-time information and power exchange.

The technology exists for the measure, monitor, and control in real time in the Smart Grid, and this technology plays an essential role in the functioning of the Smart Grid. Issues of standards, cyber security, and interoperability which are dealt with more extensively in Chapter 8 impact most definitely on communication. There is need for the formalization of the standards and protocols which will be enforced for the secured transmission of critical and highly sensitive information within the communications scheme.

Obviously, existing measuring, monitoring, and control technology will have a role in smart grid capability. Establishing appropriate standards, cyber security, and interop-

Smart Grid: Fundamentals of Design and Analysis, First Edition. James Momoh.

erability (discussed in Chapter 8) requires careful study, for example, formalizing the standards and protocols for the secure transmission of critical and highly sensitive information within the proposed communication scheme. Moreover, open architecture's plug-and-play environment will provide secure network smart sensors and control devices, control centers, protection systems, and users. Possible wired and wireless communications technologies can include:

1. Multiprotocol Label Switching (MPLS): high-performance telecommunications networks for data transmission between network nodes
2. Worldwide Interoperability for Microwave Access (WiMax): wireless telecommunication technology for point to multipoint data transmission utilizing Internet technology
3. Broadband over Power Lines (BPL): power line communication with Internet access
4. Wi-Fi: commonly used wireless local area network

Additional technologies include optical fiber, mesh, and multipoint spread spectrum.

The five characteristics of smart grid communications technology are:

1. High bandwidth
2. IP-enabled digital communication (IPv6 support is preferable)
3. Encryption
4. Cyber security
5. Support and quality of service and Voice over Internet Protocol (VoIP)

Reliable intercommunication of hardware and software will require configuring several types of network topologies. Below is a summary of the most likely candidates.

Local Area Network [5,6] consists of two or more components and high-capacity disk storage (file servers), which allow each computer in a network to access a common set of rules. LAN has operating system software which interprets input, instructs network devices, and allows users to communicate with each other. Each hardware device (computer, printer, and so on) on a LAN is a node. The LAN can operate or integrate up to several hundred computers. LAN combines high speed with a geographical spread of 1–10 km. LAN may also access other LANs or tap into Wide Area Networks. LAN with similar architectures are bridges which act as transfer points, while LAN with different architectures are gateways which convert data as it passes between systems.

LAN is a shared access technology, meaning that all of the attached devices share a common medium of communication such as coaxial, twisted pair, or fiber optics cable. A physical connection device, the Network Interface Card (NIC), connects to the network. The network software manages communication between stations on the system.

The special attributes and advantages of LAN include:

- Resource sharing: allows intelligent devices such as storage devices, programs, and data files to share resources, that is, LAN users can use the same printer on the network; the installed database and the software can be shared by multiple users
- Area covered: LAN is normally restricted to a small geographical area, for example, office building, utility, campus
- Cost and availability: application software and interface devices are affordable and off-the-shelf
- High channel speed: ability to transfer data at rates between 1 and 10 million bits per second
- Flexibility: grow/expand with low probability of error; easy to maintain and operate

LAN has three categories of data transmission:

1. Unicast transmission: a single data packet is sent from a source node to a destination (address) on the network
2. Multicast transmission: a single data packet is copied and sent to a specific subset of nodes on the network; the source node addresses the packet by using the multicast addresses
3. Broadcast transmission: a single data packet is copied and sent to all nodes on the network; the source node addresses the packet by using the broadcast address

LAN topologies define how network devices are organized. The four most common architectural structures are:

1. Bus topology: linear LAN architecture in which transmission from network station propagates the length of the medium and is received by all other stations connected to it
2. Ring bus topology: a series of devices connected to one another by unidirectional transmission links to form a single closed loop
3. Star topology: the end points on a network are connected to a common central hub or switch by dedicated links
4. Tree topology: identical to the bus topology except that branches with multiple nodes are also possible

The devices and software used in LAN utilize a standard protocol such as Ethernet/ IEEE 802.3, Token Ring/IEEE 802.5 or 880.2 (available through IEEE Press).

Home Access Network [2,3] is a LAN confined to an individual home. It enables remote control of automated digital devices and appliances throughout the house. Smart meters, smart appliances and Web-based monitoring can be integrated into this level.

Neighborhood Area Network (NAN) is a wireless community currently used for wireless local distribution applications. Ideally, it will cover an area larger than a LAN.

Some architectural structures will focus on the integration and interoperability of the various domains within the smart grid. Domains consist of groups of buildings, systems, individuals, or devices which have similar communications characteristics:

- **Bulk generation:** includes market services interface, plant control system, and generators; this domain interacts with the market operations and transmission domains through wide area networks, substation LANs, and the Internet
- **Transmission**: includes substation devices and controllers, data collectors, and electric storage; this domain interacts with bulk generation and operations through WANs and substation LANs; integrated with the distribution domain
- **Distribution:** this domain interacts with operations and customers through Field Area Networks
- **Customer:** includes customer equipment, metering, Energy Management Systems (EMS), electric storage, appliances, PHEVs, and so on
- **Service Providers:** includes utility and third party providers which handle billing customer services, and so on; this domain interacts with operations and customers primarily through the Internet
- **Operations**: includes EMS, Web Access Management System (WAMS), and SCADA; this domain can be sub-divided into ISO/RTO, transmission, and distribution
- **Market:** includes /ISOs/RTOs, aggregators, and other market participants

2.2 MONITORING, PMU, SMART METERS, AND MEASUREMENTS TECHNOLOGIES

The smart grid environment requires the upgrade of tools for sensing, metering, and measurements at all levels of the grid. These components will provide the data necessary for monitoring the grid and the power market. Sensing provides outage detection and response, evaluates the health of equipment and the integrity of the grid, eliminates meter estimations, provides energy theft protection, enables consumer choice, DSM, and various grid monitoring functions.

With regard to metering and measurement, new digital technologies using two-way communications, a variety of inputs (pricing signals, time-of-day tariff, regional transmission organization (RTO) [7] curtailments for congestion relief), a variety of outputs (real time consumption data, power quality, electric parameters), the ability to connect and disconnect, and interfaces with generators, grid operators, and customer portals to enhance power measurement. This is facilitated by the increased utilization of digital electronics for metering and measurements, advancement of the electric meter at the customer level, and installation of wide area monitoring system for advanced utility monitoring and protection [11,12]. Details of these measurements are discussed in later sections.

New digital technologies will employ two-way communication, a variety of inputs (pricing signals, time-of-day tariffs, RTO) [7] curtailments for congestion relief), a variety of outputs (real-time consumption data, power quality, electric parameters), the ability to connect and disconnect, and interfaces with generators, grid operators, and customer portals to enhance power measurement. This introduces the increased utilization of digital electronics for metering and measurements, advancement of the electric meter at the customer level, and installation of wide area monitoring systems (WAMs) for advanced utility monitoring and protection [11,12]. Details of these measurements are discussed in later sections.

2.2.1   Wide Area Monitoring Systems (WAMS)

WAMS are designed by the utilities for optimal capacity of the transmission grid and to prevent the spread of disturbances. By providing real-time information on stability and operating safety margins, WAMS give early warnings of system disturbances for the prevention and mitigation of system-wide blackouts. WAMS utilize sensors distributed throughout the network in conjunction with GPS satellites for precise time stamping of measurements in the transmission system. The integrated sensors will interface with the communication network. Phasor Measurements are a current technology that is a component of most smart grid designs.

2.2.2   Phasor Measurement Units (PMU)

Phasor Measurement Units or Synchrophasors give operators a time-stamped snapshot of the power system. The PMUs consist of bus voltage phasors and branch current phasors, in addition to information such as locations and other network parameters [9,10]. Phasor measurements are taken with high precision from different points of the power system at the same instant, allowing an operator to visualize the exact angular difference between different locations. PMUs are equipped with GPS receivers which allow synchronization of readings taken at distant points [12]. Microprocessor-based instrumentation such as protection relays and Disturbance Fault Recorders (DFRs) incorporate the PMU module with other existing functionalities as an extended feature. The IEEE standard on Synchrophasors specifies the protocol for communicating the PMU data to the Phasor Data Concentrator. Figure 2.1 illustrates the PMU measurement system from Reference 8.

PMUs ensure voltage and current with high accuracy at a rate of 2.88 kHz. They can calculate real power, reactive power, frequency, and phase angle 12 times per 60 hertz cycle. The actual sampling rate used to achieve this output is 1.4 MHz [3]. Recent trends now require fast controls and online implementations for mitigating voltage collapse in the shortest, least-cost time [8]. Over the years, researchers and engineers have found PMUs are suitable for monitoring and control of voltage stability PMUs) [6–9].

Offering wide-area situational awareness, phasor measurement work to ease congestion, bottlenecks and mitigate—or even prevent—blackouts. When integrated with Smart Grid communications technologies, the measurements taken will provide dynamic

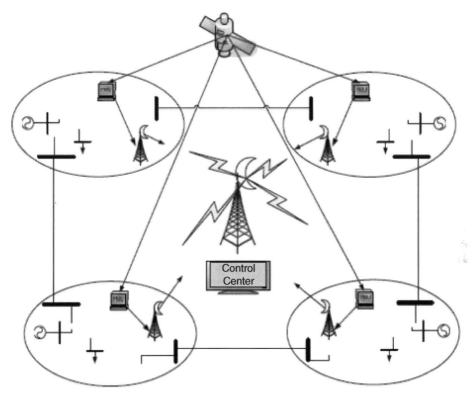

Figure 2.1. Conceptual diagram of a synchronized phasor measuring system [9].

visibility into the power system. Adoption of the Smart Grid with real time measurement will enhance every facet of the electric delivery system including generation, transmission, distribution, and consumption. It will increase the possibilities of distributed generation, bringing generation closer to those it serves.

Additional utility monitoring systems include dynamic line rating technology, conductor sensors, insulation contamination leakage current, backscatter radios technology, electronic instrument transformers, and monitors for circuit breaker, cables, batteries, temperature, and current frequency and so on [10, 11]. Research into the application and integration of these measurements into the smart grid is particularly important and continuing work.

2.2.3   Smart Meters

Smart meters have two functions: providing data on energy usage to customers (end-users) to help control cost and consumption; sending data to the utility for load factor control, peak-load requirements, and the development of pricing strategies based on consumption informationand so on Automated data reading is an additional component of both smart meters and two-way communication between customers and utilities. The development of smart meters is planned for electricity, water, and gas consumption [9].

Smart meters equip utility customers with knowledge about how much they pay per kilowatt hour and how and when they use energy. This will result in better pricing information and more accurate bills in addition to ensuring faster outage detection and restoration by the utility. Additional features will allow for demand-response rates, tax credits, tariff options, and participation in voluntary rewards programs for reduced consumption. Still other features will include remote connect/disconnect of users, appliance control and monitoring, smart thermostat, enhanced grid monitoring, switching, and prepaid metering.

With governmental assistance, large-scale deployment of smart meters has begun throughout the United States. Academic participation in the research and development of metering tools and techniques for network analysis enhancement and the use of smart meter outputs for voltage stability and security assessment and enhancement have been proposed.

2.2.4 Smart Appliances

Smart appliances cycle up and down in response to signals sent by the utility. The applicances enable customers to participate in voluntary demand response programs which award credits for limiting power use in peak demand periods or when the grid is under stress. An override function allows customers to control their appliances using the Internet.

Air conditioners, space heaters, water heaters, refrigerators, washers, and dryers represent about 20% of total electric demand during most of the day and throughout the year [10]. Grid-friendly appliances use a simple computer chip that can sense disturbances in the grid's power frequency and can turn an appliance off for a few minutes to allow the grid to stabilize during a crisis.

2.2.5 Advanced Metering Infrastructure (AMI)

AMI is the convergence of the grid, the communication infrastructure, and the supporting information infrastructure. The network-centric AMI coupled with the lack of a composite set of cross industry AMI security requirements and implementation guidance, is the primary motivation for its development. The problem domains to be addressed within AMI implementations are relatively new to the utility industry; however, precedence exists for implementing large-scale, network-centric solutions with high information assurance requirements. The defense, cable, and telecom industries offer many examples of requirements, standards, and best practices that are directly applicable to AMI implementations.

The functions of AMI can be subdivided into three major categories:

- **Market applications:** serve to reduce/eliminate labor, transportation, and infrastructure costs associated with meter reading and maintenance, increase accuracy of billing, and allow for time-based rates while reducing bad debts; facilitates informed customer participation for energy management

- **Customer applications:** serves to increase customer awareness about load reduction, reduces bad debt, and improves cash flow, and enhances customer convenience and satisfaction; provides demand response and load management to improve system reliability and performance
- **Distribution operations:** curtails customer load for grid management, optimizes network based on data collected, allows for the location of outages and restoration of service, improves customer satisfaction, reduces energy losses, improves performance in event of outage with reduced outage duration and optimization of the distribution system and distributed generation management, provides emergency demand response

An extension of AMI will be smart meters which handle customers' gas and water usage data. The issues associated with ensuring network and data security are discussed in Chapter 8.

2.3   GIS AND GOOGLE MAPPING TOOLS

GIS is useful for managing traditional electric transmission and distribution and telecom networks. It can also help to manage information about utility assets for data collection and maintenance.

Google's free downloadable Google Earth software offers geographical contextual information in an updated user-friendly platform that facilitates inquiry-based study and analysis. Users can create and share many types of dynamically-updating data over the Internet. Keyhole Markup Language (KML) allows them to overlay basic data types such as images, point data, lines, and polygons [7]. Through satellite imagery, maps are available from space to street-level. The integration of GIS with Google Earth or other mapping tools will aid in understanding the relationship of the grid network to its surroundings, for example, determining the optimal location of rights of way, placement of sensors and poles, and so on. GIS technology will provide partial context to operators and planners, for example, real-time sensors that collect the data needed to reconfigure networks for reducing outages and equipment failures.

The trends in the development of the electric power system and the expectation of future demand suggest the following needs:

1. Reducing outage time
2. Preventing power theft which causes significant unaccounted losses
3. Effective system for collection and billing system
4. Expanding services for customers
5. Effective asset management
6. Improving reliability such as SAIDI (System Average Interruption Duration Index) and SAIFI (System Average Interruption Frequency Index) for distribution networks

7. Improving analysis of customer complaint logs
8. Enhancing load flow power quality analysis and fault study for current and anticipated problems
9. Scheduling of actions such as load shedding and vegetation control

2.4 MULTIAGENT SYSTEMS (MAS) TECHNOLOGY

MAS are a computational system in which several agents cooperate to achieve a desired task. The performance of MAS can be decided by the interactions among various agents. Agents cooperate to achieve more than if they act individually.

Increasingly, MAS are the preferential choice for developing distributed systems. The development of monitoring and measurement schemes within the smart grid environment can be enhanced through the use of MAS architecture (Fig. 2.2). As an example, MAS have been utilized as a detection and diagnosis device and in system monitoring. Such architectures utilize a collection of agents such as Arbitrator Agents (AA), System Monitoring Agents (SMA), Fault Detection Agents (FDA), Diagnosis Agents (DA), a Judgment Index Agent (JIA), and a Scheduling Agent (SA). Information passes between the agents about the appropriate actions to be taken. When implemented, the process repeats itself to constantly monitor the system so that management of system conditions can be implemented instantaneously.

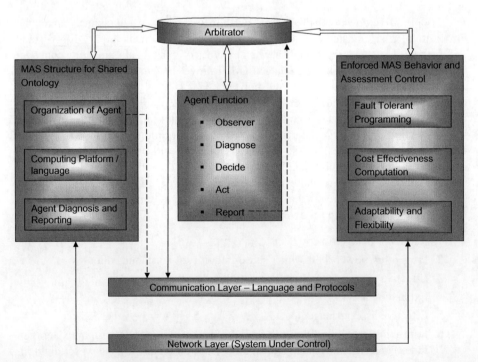

Figure 2.2. Simplified multiagent architecture.

2.4.1 Multiagent Systems for Smart Grid Implementation

As mentioned in Chapter 1, the DOE's Modern Grid Initiative [1] explains that a smart grid integrates advanced sensing technologies, control methods, and integrated communications into the present electricity grid at both transmission and distribution levels. The smart grid is expected to have the following key characteristics:

1. self-healing
2. consumer friendly
3. attack resistant
4. provide power quality for 21st-century needs
5. accommodate all generation and storage options
6. enable markets
7. optimize assets and operate efficiently

Central to the operation of any power system is its control architecture consisting of hardware and software protocols for exchanging system status and control signals. In conventional electric power systems, this is accomplished by SCADA [4, 5]. Current trends to control and monitor system operations are moving toward the use of MAS. A multiagent system is a combination of several agents working in collaboration pursuing assigned tasks to achieve the overall goal of the system. The multiagent system has become an increasingly powerful tool in developing complex systems that take advantages of agent properties: autonomy, sociality, reactivity and pro-activity [6].

The multiagent system is autonomous in that they operate without human interventions. The multiagent system is sociable in that they interact with other agents via some kind of agent communication language. The agents also perceive and react to their environment. Lastly, the multiagent system is proactive in that they are able to exhibit goal oriented behavior by taking initiatives.

2.4.2 Multiagent Specifications

In this section, the specifications of a control agent, a distributed energy resource (DER) agent, a user agent, and a database agent in the Intelligent Distributed Autonomous Power System (IDAPS) MAS are defined.

1. *Control agent:* responsibilities include monitoring system voltage and frequency to detect contingency situations or grid failures, and sending signals to the main circuit breaker to isolate the IDAPS microgrid from the utility when an upstream outage is detected; receiving electricity price ($/kWh) signal from the main grid, which may be obtained from AMI, and publishing them to the IDAPS entities
2. *DER agent:* responsibilities include storing associated DER information, monitoring and controlling DER power levels and connect/disconnect status; DER information to be stored may include DER identification number, type (solar

cells, microturbines, fuel cells), power rating (kW), local fuel availability, cost function or price at which users agree to sell, DER availability, that is, planned maintenance schedule

3. *User agent:* acts as a customer gateway that makes features of an IDAPS microgrid accessible to users; responsibilities include providing users with real-time information on entities residing in the IDAPS system; monitors electricity consumption by each critical and noncritical load; allows users to control the status of loads based on user's predefined priority

4. *Database agent:* serves as a data access point for other agents as well as users; responsibilities include storing system information, recording messages and data shared among agents.

2.4.3 Multiagent Technique

An agent of a MAS may be defined as an entity with attributes considered useful in a particular domain. In this framework, an agent is an information processor that performs autonomous actions based on information. Common agent attributes include:

- Autonomy: goal-directedness, proactive and self-starting behavior
- Collaborative behavior: the ability to work with other agents to achieve a common goal
- Knowledge-level communication ability: the ability to communicate with other agents with language resembling human speech acts rather than typical symbol-level program-to-program protocols
- Reactivity: the ability to selectively sense and act
- Temporal continuity: persistence of identity and state over long periods

MAS can be characterized by:

- Each agent has incomplete capabilities to solve a problem
- No global system control
- Decentralized data
- Asynchronous computation

For instance, the system outage of a ship could be caused by an internal system error or any external contingency from battle. To pursue the best ship performance, it is very important to restore the electric power supply as much as possible. When a fault occurs on the ship power system, the protection systems will isolate the fault from the remaining power grid. Then the system should restore the electric power to a target configuration after the outage.

An example of MAS architecture in action is a power failure on board a ship that is caused by an internal system error, an external contingency from battle, and so on.

Clearly, the goal is rapid restoration of the onboard power supply; hence, when a fault occurs, the protection system will isolate the fault, allowing the system to restore power to a target configuration after the outage.

2.5 MICROGRID AND SMART GRID COMPARISON

Research has been conducted to understand the differences between a microgrid and a smart grid. Basically, a microgrid is a local island grid that can operate as a stand-alone or as a grid-connected system. It is powered by gas turbines or renewable energy and includes special purpose inverters and a link for plug-and-play to the legacy grid. Special purpose filters overcome harmonics problems while improving power quality and efficiency. Several demonstration projects and a testbed are operating in university and government facilities. In summary, think of the microgrid as a local power provider with limited advanced control tools and the smart grid is a wide area provider with sophisticated automated decision support capabilities.

2.6 SUMMARY

This chapter has focused on various communication aspects of the smart grid. The measurement techniques described included PMUs and smart meters. GIS was introduced as a planning tool to facilitate locating important components. The relationship of MAS to the smart grid developmental process was also described.

REFERENCES

[1] J.A. Momoh. *Electric Power System Application of Optimization*. Marcel Dekker, New York, 2001.

[2] J.L. Marinho and B. Stott. "Linear Programming for Power System Network Security Applications." *IEEE Transactions on Power Apparatus and Systems* 1979, PAS-98, 837–848.

[3] R.C. Eberhart and J. Kennedy. "A New Optimizer Using Particle Swarm Theory." In *Proceedings on the Sixth International Symposium on Micromachine and Human Science* 1995, 39–31.

[4] G. Riley and J. Giarratano. *Expert Systems: Principles and Programming*. PWS Publisher, Boston, 2003.

[5] A. Englebrecht. *Computational Intelligence: An Introduction*, John Wiley & Sons, 2007.

[6] M. Dorigo and T. Stuzle. *Ant Colony Optimization*. Massachusetts Institute of Technology, Cambridge, 2004.

[7] A.G. Barto, W.B. Powell, D.C. Wunsch, and J. Si. Handbook of Learning and Approximate Dynamic Programming. *IEEE Press Series on Computational Intelligence*, 2004.

[8] J.A. Momoh. *Electrical Power System Applications of Optimization*, CRC Press, 2008.

[9] W.H. Zhange and T. Gao. "A Min-Max Method with Adaptive Weightings for Uniformly Spaced Pareto Optimum Points." *Computers and Structures* 2006, 84, 1760–1769.

[10] P.K. Skula and K. Deb. "On Finding Multiple Pareto-Optimal Solutions Using Classical and Evolutional Generating Methods." *European Journal of Operational Research* 2007, 181, 1630–1652.

[11] M. Dorigo and T. Stutzle. "The Ant Colony Optimization Metaheuristic: Algorithms, Applications and Advances." In F. Glover and G. Kochenberger, eds.: *Handboook of Metaheuristics*. Norwell, MA, Kluwer, 2002.

[12] A.G. Phadhke. "Synchronized Phasor Measurements in Power Systems." *IEEE Comput. Appl. Power* 1993, 6, 10–15.

3

PERFORMANCE ANALYSIS TOOLS FOR SMART GRID DESIGN

3.1 INTRODUCTION TO LOAD FLOW STUDIES

Load flow studies are critical to system planning and system operation. For example, data on peak load conditions assists planners in determining the size of components (conductors, transformers, reactors, and shunt capacitors), siting new generation and transmission, and planning interties with neighboring systems to meet predicted demand consistent with the North American Electric Reliability Corporation's (NERC) reliability requirements. Load flow studies identify line loads and bus voltages out of range, inappropriately large bus phase angles (and the potential for stability problems), component loads (principally transformers), proximity to Q-limits at generation buses, and other parameters having the potential to create operating difficulties. Intermediate load and off-peak (minimum) load studies are also useful, since off-peak loads can result in high voltage conditions that are not identified during peak loads. Load flow studies assist system operators in calculating power levels at each generating unit for economic dispatch, analyzing outages and other forced operating conditions (contingency analysis [7]), and coordinating power pools. In most instances, load flow studies are used to assess system performance and operations under a given condition.

Smart Grid: Fundamentals of Design and Analysis, First Edition. James Momoh.
© 2012 Institute of Electrical and Electronics Engineers. Published 2012 by John Wiley & Sons, Inc.

3.2 CHALLENGES TO LOAD FLOW IN SMART GRID AND WEAKNESSES OF THE PRESENT LOAD FLOW METHODS

Current legacy methods have weaknesses that need to be addressed prior to their use in analyzing smart grid performance and operations. Four fundamental questions should be answered:

1. What are the special features of the smart grid compared to the legacy system?
2. What computations are needed in the case of smart grid?
3. What specific directions are needed for developing a new power flow?
4. What new features of the load flow make it suitable for smart grid performance and evaluation?

Table 3.1 compares the old and desired load flow techniques.
Other features to be considered in the development of the new load flow include:

1. Condition adaptiveness [9–11] of transmission and distribution to accommodate load flows comprising renewable generation
2. Self-adaptiveness to ensure proper coordination
3. High impedance topology matching for distribution network with randomness and uncertainty requiring intelligence analytical tools
4. Since reverse power flow technique is possible, the use of FACTs devices to power electronics building blocks is essential.

Existing load flow performance tools capable of determining voltage, angle, flows, MW/MVAr, and scheduling dispatch are mostly offline although a few can give real-time results. To enhance load flow capabilities, the smart grid load flow process consists of the following steps:

1. Data acquisition for radial or mesh network
2. Existence of connection of data to assure network feasibility

TABLE 3.1. Load Flow Techniques Comparison

Old Load Flow Technique	Desired Load Flow Technique
Central generation and control	Central and distributed generation control and distributed intelligence
Load flow by Kirchhoff's laws	Load flow by electronics
Power generation according to demand	Controllable generation, fluctuating/random sources and demand in dynamic [1,6] equilibrium
Manual switching and trouble response	Automatic response and predictive avoidance
Simulation and response tracking	Monitoring overload against bottlenecks

3. Formulation of Y-bus for representing the interconnection of the system under study and determination of initial conditions
4. Solution of mismatch real and reactive powers and checking of mismatch by adjusting the initial conditions typically called the hot state
5. With the snapshot power demand (real power and reactive power demand), determining a feasible static voltage and angle to minimize mismatch

The process is defined as the iterative power simulation which is described mathematically in three different formulation frameworks. The process iterates until convergence is reached so that the program terminates with solutions of voltage, angle, flows, losses, and minimum mismatch power injected.

3.3 LOAD FLOW STATE OF THE ART: CLASSICAL, EXTENDED FORMULATIONS, AND ALGORITHMS

The traditional load flow techniques used for distribution load flow are characterized by:

1. Distribution systems are radial or weakly meshed network structures
2. High X/R ratios in the line impedances
3. Single phase loads handled by the distribution load flow program
4. Distributed Generation (DG), other renewable generation, and/or cogeneration power supplies installed in relative proximity to some load centers
5. Distribution systems with many short line segments, most of which have low impedance values

For the purpose of load flow study we model the network of buses connected by lines or switches connected to a voltage-specific source bus. Each bus may have a corresponding load composite form (consisting of inductor, shunt capacitor, or combination). The load and/or generator are connected to the buses.

The classical methods of studying load flow include:

1. Gauss–Seidal
2. Newton–Raphson
3. Fast Decouple

We summarize each of them here for easy reference.

3.3.1 Gauss–Seidal Method

This method uses Kirchhoff's current law nodal equations given as $I_{injection}$ current at the node. Suppose $I_{injection}$ = current at the node of a given connected load, then

$$I_{inj(j)} = \sum_{i=1}^{n} I_{ji}$$

where $Iinj(j)$ is the injection current at bus j and Iji = current flow from jth bus to ith bus. Rewriting, we obtain $Iinj(j) = Ybus\ Vbus$ where $Ybus$ admittance matrix is given as $Vbus$ vector of bus voltages.

If we sum the total power at a bus, the generation and load is denoted as complex power. The nonlinear load flow equation is:

$$S_{inj-k} = P_g + jQ_g - (P_{LD} + jQ_{LD})$$

$$= V_k \left(\sum_{j=1}^{n} Y_{kj} V_j \right)^*$$

This equation is solved by an iterative method for Vj if P and Q are specified. Additionally, from

$$V_i^{(k+1)} = g(V_{bus}^k)$$

$$= \frac{1}{Y_{ii}} \left(\frac{P_L^{sch} - jQ_L^{sch}}{V_L^{*(k)}} - \sum_{j=1}^{n} Y_{ij} V_j^{(k)} \right)$$

where Y_{ij} are the elements of bus admittance matrix, and P_i^{sch} and Q_i^{sch} are scheduled P and Q at each bus.

After a node voltage is updated within iteration, the new value is made available for the remaining equations within that iteration and also for the subsequent iteration. Given that the initial starting values for voltages are close to the unknown, the iterative process converges linearly.

Notably, the classical method performance is worse in a radial distribution system because of the lack of branch connections between a large set of surrounding buses. It should be noted that the injection voltage correction propagates out to the surrounding buses on the layer of neighboring buses for each iteration.

3.3.2   Newton–Raphson Method

The Newton–Raphson Method assumes an initial starting voltage use in computing mismatch power ΔS where $\Delta S = S_{ij-i}^{sch} - (V_i^{|k|})^* (\sum Y_{ij} V_j^k)$. The expression ΔS is called the mismatch power. In order to determine convergence criteria given by $\Delta S \leq \varepsilon$, where ε is a specific tolerance, accuracy index, and a sensitivity matrix is derived from the inverse Jacobian matrix of the injected power equations:

$$P_i = |V_i| \sum |Y_{ij}| |V_j| \cos(\theta_i - \theta_j - \psi_{ij})$$
$$Q_i = |V_i| \sum |Y_{ij}| |V_j| \sin(\theta_i - \theta_j - \psi_{ij})$$

where θ_i is the angle between V_i and V_j, and ψ_{ij} is admittance angle.

The complex power ΔS can be expressed in polar or rectangular form $|\Delta V| = (\Delta e + \Delta f)$ or $\Delta V = |\Delta V| \angle \theta_v$, or $\Delta S = \Delta P + \Delta Q$ respectively.

This method is excellent for large systems but does not take advantage of the radial structure of distribution and hence is computationally inefficient. The method fails when the Jacobian matrix is singular or the system becomes ill-conditioned as in the case of a low distribution X/R ratio.

3.3.3 Fast Decouple Method

The fast decouple method simplifies the Jacobian matrix by using small angle approximations to eliminate relatively small elements of the Jacobian. The method is one of the effective techniques used in power system analysis. However, it exhibits poor convergence with a high R/X ratio system. The interaction of V and θ magnitudes with active and reactive power flows cause poor convergence as well.

A variation solves current injection instead of model power injection power equations.

3.3.4 Distribution Load Flow Methods

Due to the limitation of the fast decouple method in solving an ill-conditioned system with a high X/R ratio, the distribution load flow [12] techniques above require alternative methods. We summarize the commonly used methods.

1. Forward/backward sweep methods solves branch current or load flow by using the forward sweeping method
2. Compute the nodal voltages using backward sweep approach
3. Newton method uses power mismatches at the end of feeders and laterals to iteratively solve the nodal voltage
4. Gauss method on the bus impedance matrix equation solves iteratively for the branch currents.

Method 1: Forward/Backward Sweep. This method models the distribution system as a tree network, with the slack bus denoted as the root of the tree and the branch networks as the layers which are far away from the root nodal. Weakly meshed networks are converted to a radial network by breaking the loops and injection currents computation.

The backward sweep primarily sums either the line currents or load flows from the extreme feeder (leaf) to the root. The steps of the algorithm are:

1. Select the slack bus and assume initial voltage and angle at the root, node, and other buses
2. Compute nodal current injection at the Kth iteration

$$I_i^{(k)} = \left[\frac{S_i^{sch}}{V_i^{(k-1)}} \right]^*$$

3. Start from the root with known slack bus voltages and move toward the feeder and lateral ends

4. Compute the voltage at node j

$$V_j^{k-1} = V_i^k - Z_{ij} I_{ij}^{(k)}$$

where Z_{ij} is the branch impedance between bus i and j and V_j is the latest voltage value of bus j

5. Compute the power mismatch from and check the termination criteria using

$$\Delta S_i^{(k)} = S_i^{sch} - V_i^{(k)} \left(I_i^{(k)} \right)^* \le \varepsilon$$

6. If step above is not reached repeat the previous steps until convergence is achieved.

Note that in Step 2 from each known load power S, the lateral voltages are computed or assumed. This involves V_i^{k-1} as the $k - 1$ past iteration of bus voltage and $I_i^{(k)}$ is the Kth current iteration of injected current. We do this by starting from the last branch from the lateral feeder and moving back through the tree node. This is done using the expression

$$I_i^{(k)} = \left[\frac{S_i^{sch}}{V_i^{(k)}} \right]$$

as before for all interconnected branches.

Method 2: Load Flow Based on Sensitivity Matrix for Mismatch Calculation. The distribution load flow is an improved forward/backward method utilizing a sensitivity matrix scheme to compensate the mismatch between slack bus power injection and the load flow at the feeder and lateral ends. This results in the Newton–Raphson method for distribution load flow.

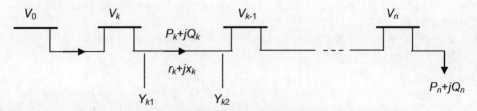

Figure 3.1. Single feeder representation.

Consider a single feeder, as shown in Figure 3.1. The steps for this method are:

1. Assume the slack bus as the root node
2. Assume P_0, Q_0 power injection at the slack bus node equal to the sum of all of the loads in the system
3. Load flows in each branch are equal to the sum of downstream connected loads. At kth iteration, start from root node with known voltage at slack bus
4. Obtain the latest V^k, P_{ij}^k, Q_{ij}^k (voltage and flows)
5. Compute power loss $= f(V^k, P_{ij}^*, Q_{ij}^*)$
6. From the loss compute receiving and power P_{ji}, Q_{ji}, and V_j
7. The loads and shunt power are taken from the received power and the remaining power is sent to the next feeder at lateral branches
8. At network solutions ΔP_L, $\Delta Q_L = 0$, when mismatch power is approximately 0; if load flow mismatch is less than the tolerance, ε, then load flow has converged
9. Update the slack bus power from the sensitivity matrix

Method 3: Bus Impedance Network. This method uses the bus impedance matrix and equivalent current injection to solve the network equation in a distribution system. It employs a simple superposition [3,4] to find the bus voltage through the system. The voltage in each bus is computed after specifying the slack bus voltage and then computing the incremental change ΔV due to current injection flowing into the network.

The steps are:

1. Assume no load system
2. Initialize the load bus voltage throughout the system using the value of the slack bus voltage
3. Modify nodal voltages due to current flow which are function of loads connected
4. The injection current is modified in Kth iteration as level changes
5. Use $I_i^{(k)} = (S_i^{sch}/V_i^{k-1})^*$ for the first equivalent current injection until getting I_j^k at the last iteration $I_0^{(k)}$
6. Compute the vector of voltage denoted as ΔV using $\Delta V_{bus}^{(k)} = Z_{bus} I_{inj}^{(k)}$, where Z_{bus} is a $\eta \times \eta$ bus impedance matrix
7. Determine the bus voltage updates throughout the network as $V_i^{k-1} = V_0 - \Delta V_i^{(k-1)}$ where V_0 is slack bus voltage at root node
8. Check mismatch power at each load bus using specified and calculated values to obtain $\Delta S = S^{spec} - \Sigma V_i^{calc} I_{ij}^{calc}$ and stop if the value of $\Delta S \leq \varepsilon$.
9. Otherwise, go to step 3.

Note that the load flow techniques for transmission or distribution system are not sufficient for smart grid load flow. These methods can easily be implemented using (i) sparsity techniques, (ii) implicit bus matrix, or (iii) computational techniques. In summary these load flow techniques are well documented in the literature. Software package are available for demonstrating their tradeoffs and compatibility.

ATC is the maximum amount of additional MW transfer possible between two parts of a power system. Additional means that existing transfers are considered part of the base case and are not included in the ATC number. Typically, these two parts are control areas or any group of power injections. Maximum refers to cases of either no overloads occurring in the system as the transfer is increased or no overloads occurring in the system *during contingencies* as the transfer is increased in online real time.

By definition, ATC is computed using the formula

$$ATC = TTC - \Sigma(CBM, TRM, \text{ and "existing TC"})$$

where the components are Total Transfer Capability (TTC), Capacity Benefit Margin (CBM), Transmission Reliability Margin (TRM), and existing Transmission Commitments.

ATC is particularly important because it signals the point where power system reliability meets electricity market efficiency. It can have a huge impact on market outcomes and system reliability, so the results of ATC tend to be of great interest to all players.

The load flow is solved using the iterate solution on

$$\begin{bmatrix} \Delta\delta \\ \Delta V \end{bmatrix} = [\mathbf{J}]^{-1} \begin{bmatrix} \Delta P \\ \Delta Q \end{bmatrix}$$

to obtain the incremental changes in the LHS vector until the terminating conditions of the power mismatch (or maximum number of iterations) is reached. After convergence, derived quantities such as losses and power factors can be computed using network equations.

A smart decoupled load flow equation could be developed based on the following assumptions.

Let

$$\delta_k - \delta_m \approx 0 \text{ such that } \cos(\delta_k - \delta_m) \approx 1 \text{ and } \sin(\delta_k - \delta_m) \approx 0$$

$$V_k \approx 1$$

$r_{km} \ll x_{km}$, which implies $g_{km} \approx 0$.
This allows a reduction of the Jacobian such that

$$\begin{bmatrix} \Delta\delta \\ \Delta V \end{bmatrix}$$

can be solved without using an iterative method to solve simultaneous equations. The approximation takes advantage of the lower sensitivities of real power with respect to (w.r.t.) voltage magnitude and reactive power w.r.t. voltage arguments.

The DC load flow equations are simply the real part of the decoupled load flow equations. This is achieved via the following additional assumptions: only angles and real power in real time are solved for by iterating

$$\Delta\boldsymbol{\delta} = [\mathbf{B}']^{-1}\Delta\mathbf{P} \text{ where } \frac{\partial\mathbf{P}}{\partial\acute{}} = \mathbf{B}' \text{ with } \frac{\partial P_k}{\partial\delta_k} = \sum_{\substack{m=1 \\ m\neq k}}^{N} b_{km} \text{ and } \frac{\partial P_k}{\partial\delta_m} = -b_{km}.$$

Recall the load flow that is solved using the iterate on

$$\begin{bmatrix} \Delta\boldsymbol{\delta} \\ \Delta\mathbf{V} \end{bmatrix} = [\mathbf{J}]^{-1}\begin{bmatrix} \Delta\mathbf{P} \\ \Delta\mathbf{Q} \end{bmatrix}$$

until the terminating conditions of the power mismatch (or maximum number of itera-tions) is reached. The solution, via any of the three approaches summarized above, can be used to compute the transfer of power from one node or set of nodes to another within a power system network. Thus, incorporating online real-time data measurement allows us to formulate the load flow for a smart grid network.

Role and Calculation of Power Transfer Distribution Factors (PTDFs). Power Transfer Distribution Factors (PTDFs) measure the sensitivity of line MW flows to a MW transfer. The line flows, from network theory, are simply functions of the voltages and angles at its terminal buses—sending and receiving. Using the chain rule, the PTDFs are represented as a function of these voltage and angle sensitivities.

The Single Linear Step ATC is extremely fast. Linearization is quite accurate in modeling the impact of contingencies and transfers. However, it only uses derivatives around the present operating point and does not model the control changes resulting from ramping up to the transfer limit. Also, special devices such as phase shifters which are generally used to redirect load flow across critical transmission interface can be included.

The possibility of generators participating in the transfer hitting limits is not modeled. Iterated Linear Step ATC considers these control changes.

3.4   CONGESTION MANAGEMENT EFFECT

Congestion is defined as a situation where load flows lead to a violation of operational limits, that is, maximal thermal loading, voltage stability, or the $n-1$ security rule. Net Transmission Capacity (NTC) is the additional transaction potential between zones A and B given a base scenario of injections.

In a meshed network, intrazonal injection variations will influence cross-border load flows and are an important factor in the determination of transmission capacities. These variations are a superposition of load flows on a given scenario of injections. Accounting for the injection variations allows for a better estimation of the transfer capacities, indifferent to the base case variations and creating thus more accurate long-term capacity index.

Due to the large number of linked equations that describe an actual electricity network, numerical methods are used to determine load flows and node voltages. The base load flow calculation method uses Newton–Raphson iterations, also called AC load flow. The computational efforts can be relieved by using decoupled load flow equations. This assumes line impedances to be particularly inductive, which is an acceptable assumption in the case of high voltage transmission lines. Going one step further, a DC load flow can be used. In this method ohmic losses are ignored by stating that lines are purely inductive. The injection of reactive power is assumed to be sufficient to keep the voltage profile in the network at a constant and nominal level. The iterative algorithm of AC load flow equations is reduced to a system of linear equations. These DC load flow equations will be used to simulate the effect of power injections on load flows in the transmission network.

$$\begin{bmatrix} ISF_{1,1} & \cdots & ISF_{1,k} \\ \vdots & \ddots & \vdots \\ ISF_{n,1} & \cdots & ISF_{n,k} \end{bmatrix} \begin{bmatrix} I_{nj1} \\ \vdots \\ I_{njk} \end{bmatrix} = \begin{bmatrix} flow_1 \\ \vdots \\ flow_n \end{bmatrix}$$

The injection in node i is generated power minus the load at node i. Injection Shift Factor represents the proportion of the power injected in node k which flows in transmission line n. The load flows to a certain reference node which is not of importance for the global system solution, but is used in the calculation of the ISFs and is necessary for a proper understanding of these factors. The difference between two ISFs gives a PTDF. A PTDF indicates the proportion of a transaction between two nodes that flows across a certain line.

3.5 LOAD FLOW FOR SMART GRID DESIGN

Load flow tools that incorporate the stochastic and random study of the smart grid could be modeled with the following implementation algorithm. Conditioning the load flow topology will require a new methodology and algorithm that will include feeders and the evolution of a time-dependent load flow. This method has been proven in terms of characteristics and usage in power system planning and operation. Hence, the interoperability of RER with smart grid specifications could account for adequate use of current methodology to perform analysis in both usual and alert states.

The implementation algorithm proposed will extend the following capability:

1. Model input of RER [2,5] and load will be changed to account for variability; the input will have to include some power distribution flow so as to advance the congested value of new estimate of P_f, Q_g, and P_d, Q_d. These attributes also have a unique load appropriate effectiveness in the performance study.
2. Sparsity may be affected because the loads of RER may be widely distributed, that is, load and size of RER has to be considered.
3. Computational challenges in new load flow with RER for smart grid that include the stochastic model may affect the independent computation.

The load flow is also used in distributed networks (see Fig. 3.2).

ATC is the maximum amount of additional MW transfer possible between two parts of a power system. Additional means that existing transfers are considered part

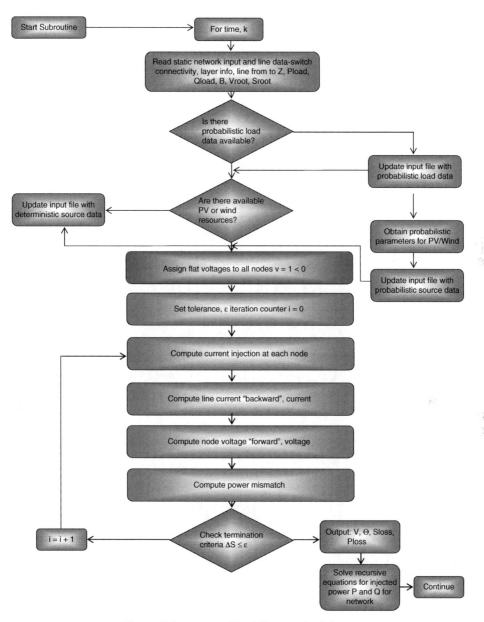

Figure 3.2. Proposed load flow methodology.

of the "base case" and are not included in the ATC number. Typically these two parts are control areas or it can be any group of power injections. The term "maximum" refers to cases of either no overloads occurring in the system as the transfer is increased or no overloads occurring in the system *during contingencies* as the transfer is increased in online real-time.

By definition, ATC is computed using the formula

$$\text{ATC} = \text{TTC} - \Sigma(\text{CBM, TRM, and "existing TC"})$$

where the components are Total Transfer Capability (TTC), Capacity Benefit Margin (CBM), Transmission Reliability Margin (TRM), and "existing Transmission Commitments".

ATC is particularly important because it signals the point where power system reliability meets electricity market efficiency. It can have a huge impact on market outcomes and system reliability; so the results of ATC are of great interest to all involved – power marketers, operators, planning engineers, and policy makers.

The load flow is solved using the iterate solution on

$$\begin{bmatrix} \Delta\boldsymbol{\delta} \\ \Delta\mathbf{V} \end{bmatrix} = [\mathbf{J}]^{-1} \begin{bmatrix} \Delta\mathbf{P} \\ \Delta\mathbf{Q} \end{bmatrix}$$

to obtain the incremental changes in the LHS vector until the terminating conditions of the power mismatch (or maximum number of iterations) is reached. After convergence, derived quantities such as losses and power factors can be computed using network equations.

A smart decoupled load flow equation could be developed based on the following assumptions.

Let

$$\delta_k - \delta_m \approx 0 \text{ such that } \cos(\delta_k - \delta_m) \approx 1 \text{ and } \sin(\delta_k - \delta_m) \approx 0$$

$$V_k \approx 1$$

$r_{km} \ll x_{km}$, which implies $g_{km} \approx 0$.
This allows a reduction of the Jacobian such that

$$\begin{bmatrix} \Delta\boldsymbol{\delta} \\ \Delta\mathbf{V} \end{bmatrix}$$

can be solved without using an iterative method to solve simultaneous equations. The approximation takes advantage of the lower sensitivities of real power with respect to (w.r.t.) voltage magnitude and reactive power w.r.t voltage arguments.

The "DC Load flow" equations are simply the real part of the decoupled load flow equations. This is achieved via the following additional assumptions:
Only angles and real power in real time are solved for by iterating

$$\Delta\delta = [\mathbf{B}']^{-1}\Delta\mathbf{P} \text{ where } \frac{\partial\mathbf{P}}{\partial\delta} = \mathbf{B}' \text{ with } \frac{\partial P_k}{\partial\delta_k} = \sum_{\substack{m=1 \\ m\neq k}}^{N} b_{km} \text{ and } \frac{\partial P_k}{\partial\delta_m} = -b_{km}.$$

Recall the load flow that was solved using the iterate on

$$\begin{bmatrix} \Delta\delta \\ \Delta\mathbf{V} \end{bmatrix} = [\mathbf{J}]^{-1}\begin{bmatrix} \Delta\mathbf{P} \\ \Delta\mathbf{Q} \end{bmatrix}$$

until the terminating conditions of the power mismatch (or maximum number of iterations) is reached. The solution, via any of the three approaches summarized above, can be used to compute the transfer of power from one node or set of nodes to another within a power system network. Thus, by incorporating online real-time data measurement, we could formulate the load flow for a smart grid network.

3.5.1 Cases for the Development of Stochastic Dynamic Optimal Power Flow (DSOPF)

DSOPF is being developed by the author. The DSOPF computational algorithm [8] has the following built-in performance measures that are also defined for other general purpose tools:

- Controllability and interoperability: This is important for enabling different devices, systems, and subsystems to provide greater observability when different devices interact as agents for cooperation and benefits.
- Reliability: quality measure of electricity delivered to achieve adequacy and performance using intelligence tools, support devices, and software; ability to achieve power quality and improve voltage profile is one of the attributes of the smart grid
- Adaptability and sustainability: ability of the grid to adapt to changes; meeting energy needs in a way that can sustain life and civilization
- Anticipatory behavior and affirmation of security: ability of the grid to anticipate different scenarios and prepare to handle the dynamic changes while guaranteeing system security.

3.6 DSOPF APPLICATION TO THE SMART GRID

Adaptive Dynamic Programming (ADP) is a computational intelligence technique that incorporates time Framework for Implementation of DSOPF [1,2].

There is a need for a generalized framework for solving the many classes of power system problems where programmers, domain experts, and so on, can submit their challenge problem. The collective knowledge will published and posted on the Web for dissemination. Figure 3.3 shows the general framework for the application of ADP to develop a new class of OPF problems called DSOPF; it is divided into three modules.

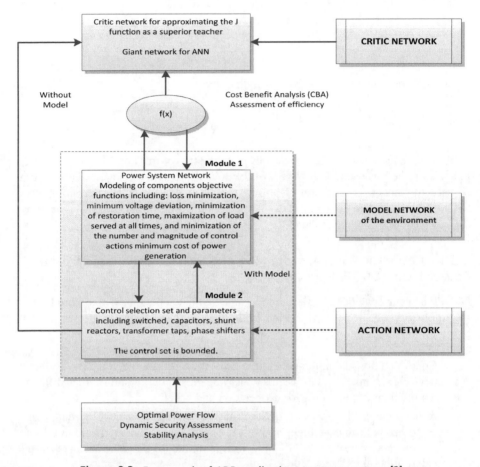

Figure 3.3. Framework of ADP applications to power systems [3].

Module 1: Read power system parameters and obtain distribution function for state estimation of measurement errors inherent in data; ascertain and improve accuracy of data. Infer relationships between the past data and future of unknown period using time series and dynamical systems; in all cases determine the time-dependent model approximation behavior of the generation data. Define the model with the uncertainties, including defining the problem objective and constraint functions for each problem.

Module 2: Determine the feasibility region of operation of the power system and the emergency state with corresponding violations under different contingencies. Enumerate and schedule different control options over time for the different contingency scenarios. Coordinate the controls and perform post optimizations of additional changes. Evaluate results and perform sensitivity analysis studies.

Module 3: Address the postoptimization process through cost benefit analysis to evaluate the various controls (cost effectiveness and efficiency). In the power system parlance, a big network, which will perform this evaluation, is essential and indispensable. The critic network from ADP techniques will help realize the dual goals of cost effectiveness and efficiency of the solution via the optimization process.

3.7 STATIC SECURITY ASSESSMENT (SSA) AND CONTINGENCIES

System security refers to the ability of the power system to withstand probable disturbance with minimal disruption of service. In an operational environment, security assessment involves predicting the vulnerability of the system to possible disruptive events in real time.

Actual operating conditions change constantly because of maintenance requirements, forced outages, and load patterns. The important options available to improve upon an insecure condition include starting an available unit, rescheduling generation, or asking for assistance from a neighboring system. The concept of DyLiacco's security-state diagram, shown in Figure 3.4, shows the principal operating states:

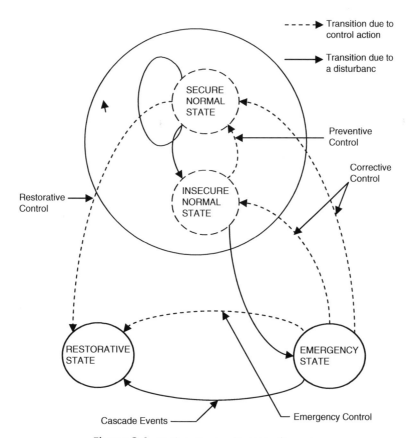

Figure 3.4. DyLiacco's security-state diagram.

1. Secure or normal state: all system loads are satisfied at the specified voltage levels
2. Emergency state: some operating limits are violated, for example, overloaded lines
3. Restorative state: some loads are not met, that is, partial or total blackout, but the operating portion of the system is in a normal state.

Assume that a transmission line sustains a fault, which causes an outage. This will result in a redistribution of power flows and changes in all voltages in the system. If this redistribution results in a normal system condition then the predisturbance state was both normal and secure relative to this event. If an emergency condition occurred, then the predisturbance state was normal, but insecure. Similarly, a system in the emergency state can be forced to return to normal following some corrective control measures. However, depending on the severity of the emergency, loads may be shed to alleviate a more catastrophic situation leading to a partially normal system (i.e., restorative state). System restoration involves activities to restore service to all interrupted loads.

Steady-state security involves situations where the transients following a disturbance have decayed, but where some limit violations could not be tolerated for long. The loss of a transmission link, for example, after the transients have died out, may result in an overloaded line, or an over-voltage condition. The system may tolerate such limit violations for a short period. During such a period corrective action should be taken. If corrective action is not possible, then the predisturbance state is seriously insecure and some preventive measures should be carried out. Analysis tools required address steady-state operation, that is, load flow and related analysis methods.

In Figure 3.5 a block diagram illustrates the functions associated with on-line security analysis. The output of the state estimator can be used directly to determine the security state (normal or emergency). For an emergency state, the next step is to specify the required corrective action and apply it before it is too late. For a normal state, it is not usually known if a postulated disturbance will or will not cause an emergency. As a result, contingency analysis is carried out using three data sources: the precontingency state (state estimator output); a model of the external system; and a specified list of contingencies. The results of contingency analysis are reevaluated further to examine the level of system security. This is referred to as security analysis. The outcome of security analysis will yield information on what to do subsequently. If the system is deemed secure, then nothing is done till the next cycle of analysis (30 minutes or an hour later). If the system is deemed insecure, then preventive measures are evaluated but not necessarily carried out. At this stage the system operator will exercise some judgment over the desirability of preventive action since, invariably, it may lead to less favorable operational economics.

3.8 CONTINGENCIES AND THEIR CLASSIFICATION

Steady-state contingency analysis predicts power flows and bus voltage conditions following events such as line outages, transformer outages, and generator outages.

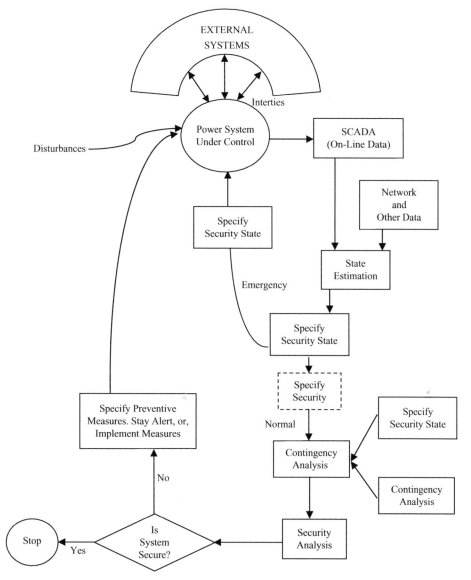

Figure 3.5. System security and associated functions.

There are many reasons for transmission line and transformer outages. The simplest reason is planned maintenance. Another involves switching operations to control power flows in the network and/or to overcome voltage problems. In either case, the outages are caused by operators performing daily dispatching and maintenance functions. Another class of outages is called forced outages, for example, a line that has experienced a permanent fault is automatically de-energized by circuit breakers, or an overloaded line or transformer is de-energized to protect it from damage. In either case, the operator needs to know the effects of the outage on power flows and voltage conditions throughout the system in order to take preventive measures before the outages occur. In the case of planned maintenance, the operator will require a fore-casted load flow case at the time of planned outage. In the case of switching operations for flow and voltage control, the operator will require a load flow solution for the present preoutage condition.

Forced outage cases are more complicated since they occur with very low proba-bilities that are time- and weather-dependent. Usually, single-line outages are more probable than double or multiple outages, but double and multiple outages occasionally occur during severe weather. As a result of the large amount of computations involved, only single-line contingencies are considered. Generator outages occur for reasons similar to those of line and transformer outages. Power plants are taken off-line for operational and maintenance requirements. There are also forced generator outages caused by equipment failures, line faults near the generator, and so on.

3.8.1 Steady-State Contingency Analysis

Steady-state contingency analysis predicts power flows and bus voltage conditions fol-lowing transmission line outages, transformer outages, and generator outages. Models of Contingency Evaluation

In transmission line or transformer outage contingencies it is assumed that input/demand load flow variables will not change due to the outage. Therefore, specified real and reactive loads, together with real generation and generator bus voltages, will be constant before and after the outage. Clearly, this line-outage model is only approxi-mate. The loss of a major transmission line will cause changes in voltage and power flow conditions. The first consequence is that power system losses will change. In the above model, these losses will be accounted for by a changed slack-bus generation level. Changes in voltage conditions at load busses normally mean changes in the load itself, especially when it is represented as an impedance load. Only with an adequate load model for each bus can the accuracy of contingency analysis be improved. For most purposes, the errors introduced by the above modeling assumption are fewer than those resulting from inaccuracies in the input data.

The case of a generator outage is more complicated for a number of reasons. Immediately following the outage (first few seconds), the remainder of the generating system will be unable to respond to the resulting generation/load unbalance by increas-ing its generation level. System frequency will drop, with the net effect being an overall and fairly uniform reduction in system load because loads are frequency-dependent and tending to decrease or increase with corresponding decreases and increases in fre-quency. Because of this drop in frequency, as well as the serious violation of the requirements on scheduled net power flow interchanges with neighboring systems, generation levels of various generators will be automatically controlled to restore both normal system frequency and net interchanges with neighboring systems through auto-matic generation control (AGC) within a few minutes. The use of economic dispatch allows the desired optimal generation levels to be accomplished, although they may not be economically and/or environmentally optimal. This final steady-state condition is reached in several minutes after the disturbance.

In practice, many smart grid contingencies will require the same number of fast load flow techniques for performing contingency analysis. In all of these techniques, the starting point is that of the preoutage solution. Some of these techniques are discussed below.

3.8.2 Performance Indices

A security-type performance index ranks the severity of various contingencies. The following index is an illustration:

$$J = \frac{1}{2} \sum_k (V_k - V_{k-ref})^2 W_k$$

where V_k is the magnitude of the voltage at bus, V_{k-ref} is the nominal (usually rated) voltage at bus k, and W_k is the weighting number associated with the relative importance, or the allowable range of voltage variation, at bus k. For a precontingency condition, J will take the value of J_0. For the ith contingency J will take the value of J_1. The contingencies are then ranked according to the J_1 values. Because of the quadratic nature of J_1, more severe contingencies will result in higher J_1 values.

A similar performance index can be formulated for line power flow limit violations—for example,:

$$J' = \frac{1}{2} \sum_k W_k \left(\frac{T_k}{T_{k\,max}} \right)^2$$

where T_k is the real power flow on line k, $T_{k\,max}$ is the corresponding maximum power flow limit, and W_k is a weighting factor.

External System Equivalents. In the online control and operational context, the system being controlled is usually interconnected to other systems. Normally, a contingency in one system will have the highest repercussions within that system. There are always cases, however, where a contingency in one system is strongly felt in another, for example, the loss of a major generating unit may cause power flow limit violations elsewhere. The difficulty in predicting the impact of a contingency arises from the fact that the external network is not monitored as carefully as the internal network. Through state estimation, all internal system voltage magnitudes and angles, power flows, generations, load, and network topology are known online. As for the external system, online information is normally restricted to items such as intertie power flows, status of major lines and generators, and possibly individual unit outputs.

When searching for an exact power flow solution to a postulated contingency, the state of the entire network (internal and external) should be known to establish the precontingency base case. Since the state of the external network is not fully known, some approximations are required. In this section, two types of approximations are considered. The first is based on sensitivity approaches, and the second on network reduction approaches. These will be illustrated, following some important definitions of terms.

The starting point is the so-called *base case* load flow solution. In a planning study this can be a complete network solution for a given set of loads, generation, and network configuration. In the case of online operations, it can be a solution of the internal plus the boundary system obtained from on-line state estimation using live measurement data.

In performing studies on the internal system, it is possible to deviate from the base case to account for line outage or generator contingencies. Normally, the existing classification of boundary nodes are retained, but there are instances when a load boundary bus may be reclassified as a generation bus. For example, if the boundary bus is close to a strong external system generation center, its voltage level will be maintained at a constant value, or nearly so, because of the external influence. In fact, some operators have reclassified all boundary load busses as generation busses, claiming higher accuracy of solutions.

The steps in the utilization of network equivalents are:

1. Perform network reduction using an appropriate technique
2. Given the base case (preoutage) solution, compute boundary bus injections
3. Reclassify some boundary busses as generation busses if necessary
4. Postulate a contingency list
5. For each contingency in the list, solve the load flow problem using a fast contingency evaluation technique that is initiated by the base case solution

3.8.3   Sensitivity-Based Approaches

As shown above, the external equivalent network depends on real and reactive external bus injections and voltage magnitudes. Following a line outage or generator contingency, the external voltages, and possibly some injections will be different. This means that a postoutage equivalent is required. Obviously, it defeats the entire purpose of network reduction. A more crucial problem is that external system injections and voltages are actually unknown in an online environment..

There are two possible solution paths. In the first path, during offline studies, it is important to obtain equivalent networks that are insensitive to external conditions and to internal outages. If this objective is achieved, the next step uses online measurements to calibrate the parametric values of equivalent lines and injections.

3.9   CONTINGENCY STUDIES FOR THE SMART GRID

Contingency studies are proposed planning/operation tool for assessing the impact of unit or line outage in an integrated smart grid environment. This could be a single or multiple line outages called $N - 1, N - 2, N - 3, \ldots$ contingency. There are two types of contingency:

1. AC automatic contingency screening/filtering
2. AC automatic contingency control

Use of the contingency set includes:

1. These contingencies sets are used in SSA as well as in DSA. The SSA consists of the following elements.
   a. Load flow base case
   b. Schedule contingency
   c. Develop and model a PI for each contingency
2. Provide the studies for violation check against a given threshold, PI ≤ Threshold for violated contingency case in ascending order
3. Develop security measure to aid improvement in recommendation and display

SSA is well known for classical power system. The weak points are: includes

1. Selection of weighting measure
2. Exponent factors
3. The probability of selecting a contingency
4. Lack of human intelligence and relative information on knowledge-base for decision-making by nonexperts. Work by the author has been planned using ES, ANN, and fuzzy sets. These schemes do not include time against sources, variability in selecting parameters for contingency studies.

The author proposes an ACS capable of including:

1. RER with variability in output power as contingency
2. Loss of measured unit needed for performance study
3. Probabilistic measure of contingency to be added to the input
4. The resulting PI index needs to be replaced with time stamp measurements using PI- OSIsoft to predict, analyze, and view recommendations by classified behavior and impact of contingency under study in real time, hence creating a new opportunity in research.

There are many opportunities for designing new computational algorithms to support the deployment of a smart grid equipped with advanced tools for system security, reliability, sustainability, and affordability.

3.10   SUMMARY

This chapter has discussed several needed tools for analysis of smart grid design, operation, and performance. Such tools include load flow, optimal power flow, and static security assessments and contingencies. Reviews of classical methods were presented along with details about each solution tool's incorporation into the smart grid.

REFERENCES

[1] M.H. Mickle and T.W. Sze. *Optimization in Systems Engineering*. Scranton, 1972.

[2] M.E. El-Hawary. *Electrical Power Systems Design and Analysis*. IEEE Press Series on Power Engineering, 2008.

[3] J.A. Momoh. *Electric Power System Application of Optimization*. Marcel Dekker, New York, 2001.

[4] J.L. Marinho and B. Stott. "Linear Programming for Power System Network Security Applications." *IEEE Transactions on Power Apparatus and Systems* 1979, PAS-98837–848.

[5] R.C. Eberhart and J. Kennedy. "A New Optimizer Using Particle Swarm Theory." In *Proceedings on the Sixth International Symposium on Micromachine and Human Science*, 1995, 39–31.

[6] A.G. Barto, W.B. Powell, D.C. Wunsch, and J. Si. *Handbook of Learning and Approximate Dynamic Programming*. IEEE Press Series on Computational Intelligence, 2004.

[7] J. Momoh. *Electrical Power System Applications of Optimization*. CRC Press, 2008.

[8] L. Zhao and A. Abur. "Multiarea State Estimation Using Synchronized Phasor Measurements." *IEEE Transactions on Power Systems* 2005, 20, 611–617.

[9] "Appendix B2: A Systems View of the Modern Grid-Sensing and Measurement." *National Energy Technology Laboratory*, 2007.

[10] T. Bottorff. "PG&E Smart Meter: Smart Meter Program," NARUC Summer Meeting 2007.

[11] D. Zhengchun, N. Zhenyong, and F. Wanliang. "Block QR decomposition based Power System State Estimation Algorithm." *ScienceDirect* 2005.

[12] M.S. Srinivas. "Distribution Load Flows: A Brief Review." IEEE Power Engineering Winter Meeting 2000. Vol. 2, pp. 942–945. August 2002.

SUGGESTED READINGS

M. Dorigo and T. Stutzle. "The Ant Colony Optimization Metaheuristic: Algorithms, Applications and Advances." In F. Glover and G. Kochenberger, eds.: *Handboook of Metaheuristics*. Norwell, MA, Kluwer, 2002.

M. Dorigo and T. Stuzle. *Ant Colony Optimization*. Massachusetts Institute of Technology, Cambridge, 2004.

A. Englebrecht. *Computational Intelligence: An Introduction*. John Wiley & Sons, 2007.

B. Milosevic and M. Begovic. "Voltage-Stability Protection and Control Using a Wide-Area Network of Phasor Measurements." *IEEE Transactions on Power Systems* 2003, 18, 121–127.

A.G. Phadhke. "Synchronized Phasor Measurments in Power Systems." *IEEE Computer Applications in Power* 1993, 6, 10–15.

G. Riley and J. Giarratano. *Expert Systems: Principles and Programming*. PWS Publisher, Boston, 2003.

P.K. Skula and K. Deb. "On Finding Multiple Pareto-Optimal Solutions Using Classical and Evolutional Generating Methods." *European Journal of Operational Research* 2007, 181, 1630–1652.

C.W. Taylor. "The Future in On-Line Security Assessment and Wide-Area Stability Control." *IEEE Power Engineering Society* 2000, 1, 78–83.

W.H. Zhange and T. Gao. "A Min-Max Method with Adaptive Weightings for Uniformly Spaced Pareto Optimum Points." *Computers and Structures* 2006, 84, 1760–1769.

4

STABILITY ANALYSIS TOOLS FOR SMART GRID

4.1 INTRODUCTION TO STABILITY

As electric power networks worldwide expand to accommodate more generation, RER, and control devices, the physical and technical consequences resulting from weaknesses in generation, transmission, and distribution will become more costly to society and the environment. To date, analysis tools that provide operational and planning guidance have been limited to studying systems with static and dynamic models of generators, busses, controls (exciters and governors), and FACTs devices. Little work has been undertaken to upgrade these tools to include predictability and adaptiveness, and to provide solutions for managing the power system of the future. Thus, this chapter first reviews the extant research in stability assessment, and follows with a proposed framework for the design of the new tools needed for smart grid analysis and design. The concept and techniques used in voltage stability assessment are described next.

4.2 STRENGTHS AND WEAKNESSES OF EXISTING VOLTAGE STABILITY ANALYSIS TOOLS

The characteristics listed here are inherent in the analytic tools for the smart grid, yet are not incorporated in the analytic tools for the existing power system network:

Smart Grid: Fundamentals of Design and Analysis, First Edition. James Momoh.
© 2012 Institute of Electrical and Electronics Engineers. Published 2012 by John Wiley & Sons, Inc.

- Robustness: Persistence of a system's characteristic behavior under perturbations or conditions of uncertainty.
- Scalability: Ability of a system, network, or process to accommodate growing amounts of work; ability to be enlarged to accommodate growth.
- Stochasticity: Time development (be it deterministic or essentially probabilistic) that is analyzable in terms of probability.
- Predictivity: Rigorous, often quantitative, forecast of what will occur under specific conditions.
- Adaptability: System can adapt its behavior according to changes in its environment or in parts of the system itself.
- Online real-time data acquisition: Instantaneous acquisition of data.

Other suitable tools and techniques include:

WAM techniques: measurements of voltage, angle, frequency, control series, and available resources for load state conditions; data is usually assumed or computed for static model with advent of GPS importance; new advances in the design of PMUs, smart meters, state estimation (SE), and FIDR monitor and control data for assessing stability, mitigating volatility, and achieving high-order efficiency and reliability.

Phasor Measurement Unit: PMUs are high-speed, time-synchronized digital recorders that measure voltage, current, and frequency on the transmission system, and calculate voltage and current magnitudes, phase angles, and real and reactive power flows. PMU data can be applied to the following:

1. Asset management
2. Voltage stability
3. Angle stability assessment
4. Designing optimum controls

Smart Meters: Two-way electronic communication meter or other device measuring electricity, natural gas, or water consumption.

Similar meters, usually referred to as interval or time-of-use meters, have existed for years, but smart meters usually involve a real-time or near real-time sensors, power outage notification, and power quality monitoring. These additional features are more than simple automated meter reading (AMR). They are similar in many respects to Advanced Metering Infrastructure (AMI) meters. Smart meters are also believed to be a less costly alternative to traditional interval or time-of-use meters and are intended to be used on a wide scale with all customer classes, including residential customers. Interval and time-of-use meters are more of a legacy technology that historically have been installed to measure commercial and industrial customers, but typically provide no AMR functionality. Smart meters may be part of a smart grid, but alone do not constitute a smart grid.

Design of Architecture for Smart Grid with RER: The legacy system is built around central generation power sources and static load models. We propose a new expandable model schematic of a smart grid with an RER single line diagram.

Defining stability methods using these conditions requires research efforts that encompass:

1. Voltage stability: Use of probability distribution function to model the resources (wind, solar, and so on) that will provide the average power output included in the network model; load models are measured using stochastic, time series methods; line voltages are provided with some probabilistic measure to assure their capture. The steps are:

2. Determine the operating point for stability assessment by performing a probabilistic load flow which starts from the base case of load flow for prime loading condition

3. Define or rank contingency based on model of choice for which selection methods include:

    a) Real-time usage of ant colony

    b) ANN

    c) Mixture to increase the selection of condition

    d) Perform the screening evaluation using one of the new index methods

    e) Determine distance from voltage collapse

These points include the collaboration of model and sensitivity analysis methods to determine the equilibrium point and to incorporate control measures to mitigate stability methods for the different scenarios of events studied.

To enhance the algorithm for smart grid, we further include a real time voltage index which is based on phasor measurement data as a means of state estimation. This involves a modification of the step above, that is, computation of the new state of the system with respect to RER and use of the data from base case load flow for different time scales to compute new voltage stability based on RER data.

The results of stability condition here require a stochastic central approach compared with the classical method of mitigation, unlike offline study with the improved system for enhancing dispatching in study mode. It can be run at fixed intervals as it can provide the most recent estimate of system conditions.

It can also automatically trigger computational engine of fast steady state. The output or input data use different scenarios which are available for developing visualization tools. The revised stability tool for smart grid will form the basis of new software for next generation of SCADA or EMS systems.

The stability result margin concept is a simple way to determine the current or actual system condition from the critical state in which a source perturbation can cause instability in the smart grid environment. This will be important so that communication infrastructure can account for necessary corrective and preventive actions. The indicators generally used are:

$$\text{Stability power flow} = 100\% * (P_{max} - P_{base}) / P_{base}$$

$$\text{Stability rescue voltage} = 100\% * (V_{max} - V_{base}) / V_{base}$$

TABLE 4.1. Old and New Grid Methodology

Methodology	Old Grid	New Grid
Model load	Static	Dynamic
Resources	Deterministic	Stochastic
FACTS devices and controls	Specified	Adaptive
Risk management	Deterministic	Random
Protection platform	Defined	Adaptive

Load base Voltage Stability Communication Point (VSCP): Based on required successive load changes and delivery of voltage desired to determine the MW/MVAr that could lead to voltage collapse

These techniques could be used to solve the classical voltage stability of old networks. Because of the new features, as discussed in Table 4.1, we need to define the tools needed to achieve this ultimate simulation platform and challenges.

The various indications of instability to be determined are meant to rank sensitivity of stability. Simply, they measure maximum transfer capability across the interfacing region in an interconnected transmission line. This is particularly useful for the smart grid where distributed resources, RER, and controls are widely spread.

Transient stability studies are needed in this case and optimization of RER and other resources to ensure that stability is derived. We propose the algorithm in Figure 4.1 and enumerate the steps to be taken in realizing this:

1. Analyze and define the system configuration with all networks in service
2. Simulate faults such as single-line-to-ground, three phase, or server contingency study
3. Evaluate the impact of contingency for branch, unit outage with voltage stability criteria
4. Perform stability analysis for voltage and angle.

The nonlinear system model used for transient stability analysis describes the system by a set of differential equations and a set of algebraic equations. Generally, the differential equations are machine equations, control system equations, and so on, and the algebraic equations are system voltage equations involving the network admittance matrix. The time simulation method and direct method are often used for transient stability analysis. The first method determines transient stability by solving the system differential equation step by step, while the second method determines the system transient stability without explicitly solving the system differential equations. This approach is appealing and has received considerable attention. Energy-based methods are a special case of the more general Lyapunov's second or direct method, the energy function being the possible Lyapunov function.

The challenges of implementation are:

1. They do not include time stamp of load change
2. They do not account for real-time information on system topology and devices contributing to system changes

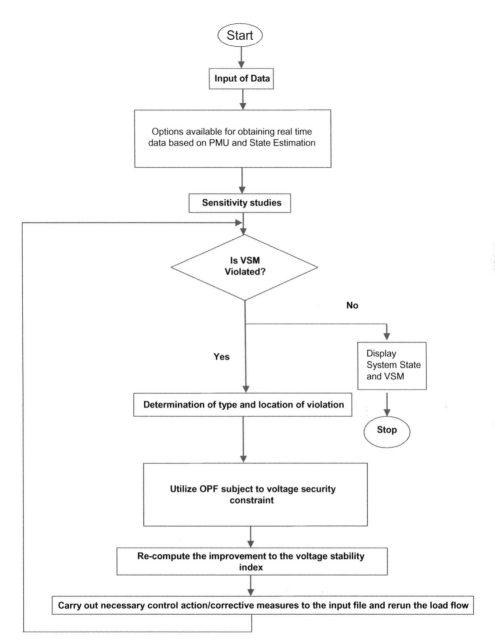

Figure 4.1. Flowchart of voltage stability and PMU that enhances the current grid.

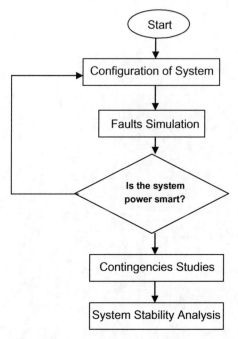

Figure 4.2. Voltage stability study algorithm.

3. They do not inform the decision-maker of real-time status (location of system parameters)

We propose that future and ongoing algorithms should include:

1. SE or PMU data filtering so that appropriate data on system parameters
2. A practical index for voltage stability based on input data or estimated system parameters
3. Assessment (preventive/corrective) scheme to guide the decision-maker
4. OPF formulation to overcome the collapse or instability

Note that this scheme is useful for both bulk power systems and distribution networks which require stability.

4.3 VOLTAGE STABILITY ASSESSMENT

Voltage stability has many definitions. It is a fast phenomenon for engineers involved with the operation of induction motors, air conditioning loads, or HVDC links, while it is a slow phenomenon (involving, for example, mechanical tap changing) for others. Voltage instability or collapse is a dynamic process. The term stability as used here implies that a dynamic system is being discussed. A power system is a dynamic system. In contrast to rotor angle (synchronous) stability, the dynamics involve mainly the loads and the means for voltage control. Voltage stability is alternatively called load stability.

The loss of lines or generators can sometimes cause voltage quality degradation, a phenomenon that is attributed to lack of sufficient reactive reserve when the power system experiences a heavy load or a severe contingency. Thus, voltage instability is characterized in such a way that voltage magnitude of the power system decreases gradually and then rapidly as the system nears the collapsing point. Voltage stability is classified as either static voltage stability or dynamic voltage stability. The latter is further classified into small signal stability and large disturbance stability problems.

In dynamic voltage stability analysis, exact models of transformers, SVCs, induction motors, and other types of loads are usually included in problem formulations in addition to models of generators, exciters, and other controllers. Small signal voltage stability problems are formulated as a combination of differential and algebraic equations that are linearized about an equilibrium point. eigenvalue analysis methods are used to analyze system dynamic behavior. Small signal analysis can provide useful information on modes of voltage instability and is instructive in locating VAR compensations and designing controllers. Large disturbance voltage stability is mainly approached using numerical simulation techniques. Since system dynamics are described by nonlinear differential and algebraic equations that cannot be linearized in nature, the mechanism of voltage collapse may be explained as a saddle node bifurcation. Voltage collapse is analyzed based on a center manifold voltage collapse model.

4.3.1 Voltage Stability and Voltage Collapse

Voltage stability has often been viewed as a steady-state viability problem suitable for static analysis techniques. The ability to transfer reactive power from production sources to consumption sinks during steady operating conditions is a major aspect of voltage stability. A power system in a given operating state and subject to a given disturbance undergoes voltage collapse if postdisturbance equilibrium voltages are prespecified acceptable limits in a significant part of the system. Voltage collapse may be total (blackout) or partial.

Assume that a power system undergoes a sudden increase of reactive power demand following a system contingency; the additional demand is met by the reactive power reserves carried by the generators and compensators. Because of a combination of events and system conditions it is possible that the additional reactive power demand may lead to voltage collapse in part or all of the system. A typical sequence of events leading to a voltage collapse is:

- The power system is experiencing abnormal operating conditions with large generating units near the load centers being out of service. Some EHV lines are heavily loaded and reactive power resources are low.
- A heavily loaded line is lost which causes additional loading on the remaining adjacent lines. This increases the reactive power losses in the lines (Q absorbed

by a line increases rapidly for loads above surge impedance loading), causing a heavy reactive power demand on the system.

- Immediately following the loss of the line, a considerable voltage reduction occurs at adjacent load centers due to the extra reactive power demand. This causes a load reduction, and the resulting reduction in power flow through the lines should have a stabilizing effect. However, the generator AVRs quickly resolve terminal voltages by increasing excitation. The resulting additional reactive power flow through the inductances associated with generator transformers and lines increases the voltage drop across each of these elements.

At this stage, generators are likely to be within the limits of P-Q output capabilities, that is, within the armature and field current heating limits. The speed governors regulate the frequency by reducing MW output.

- The EHV level voltage reduction at load centers is reflected in the distribution system. The under load tap changers (ULTCs) of substation transformers restore distribution voltages and loads to prefault levels in about 2 to 4 minutes. With each tap change operation, the resulting increment in load on EHV lines increases the line XI^2 and RI^2 losses, which in turn cause greater drops in the EHV lines. If an EHV line is loaded considerably above the SIL, each MVA increase in line flow would cause several MVArs of line losses.
- With each tap-changing operation, the reactive output of generators throughout the system increases. One by one, generators gradually exceed their reactive power capability limits (imposed by maximum allowable continuous field current). When the first generator reaches its field current limit, its terminal voltage drops. At the reduced terminal voltage for a fixed MW output, the armature current increases, and might further limit reactive output to keep the armature current within allowable limits. Its share of reactive loading transfers to other generators and causes more overloading. With fewer generators on automatic excitation control, the system is prone to voltage instability that is likely compounded by the reduced effectiveness of shunt compensators at low voltages. Eventually, there is voltage collapse or cascading, possibly leading to loss of generation synchronism.

Voltage security is the ability of a system to operate in a stable mode and to remain stable following credible contingencies or load increases. It often means that the existence of considerable margin from an operating point to the voltage instability point involves credible contingencies.

4.3.2 Classification of Voltage Stability

The three categories of voltage problems are:

1. Primary phenomena related to system structure: Reflect the autonomous response of the system to reactive supply/demand imbalances.

2. Secondary phenomena related to control actions: Reflect the counterproductive nature of some manual or automatic control actions.

3. Tertiary phenomena resulting from interaction of the above.

This classification implies that the problems involve both static and dynamic aspects of system components. Voltage collapse dynamics span a range in time from a fraction of a second to tens of minutes. Time frame charts are used to describe dynamic phenomena which show time responses from equipment that may affect voltage stability. The time frames could be considered to become very fast transient, transient, and long term. The main characteristics of the three time frames are:

1. Very fast transient voltage collapse: Involves network RLC components having very fast response; the time range is from microseconds to milliseconds.

2. Transient voltage collapse: Involves a large disturbance and loads having a rapid response; motor dynamics following a fault are often the main concern; the time frame is one to several seconds.

3. Long-term voltage collapse: Involving a load increase or a power transfer increase; shows load restoration by tap-changer and generator current limiting; manual actions by system operators may be important; the time frame is usually 0.5 to 30 minutes.

Since voltage stability is affected by various system components in a wide time range, in order to tackle this problem, one must consider proper modeling and analysis methods. Currently, voltage stability approaches mainly include static and dynamic, that is, transient voltage collapse and long-term voltage collapse.

4.3.3 Static Stability (Type I Instability)

Suppose that $\Delta \dot{x}_S = \Delta \dot{x}_F = 0$. Then we have a static situation with all equations being algebraic. Let all the voltage deviations in $\Delta \hat{V}_g$ and $\Delta \hat{V}_l$ be denoted by $\Delta \hat{V}$, then the rest of the algebraic variables can be eliminated (assuming constant power load) to express $\Delta \hat{V} = J_s^{-1} H \Delta p_L$. If $\det(J_s) \to 0$ as load is increased it is called a Type I static instability, that is, the system cannot handle increased load.

4.3.4 Dynamic Stability (Type II Instability)

Eliminating the algebraic variables and assuming $\Delta u \equiv 0$,

$$\begin{bmatrix} \Delta \dot{x}_S \\ \Delta \dot{x}_F \end{bmatrix} = \begin{bmatrix} A_{SS} & A_{SF} \\ A_{FS} & A_{FF} \end{bmatrix} \begin{bmatrix} \Delta x_S \\ \Delta x_F \end{bmatrix} + [H_1] \Delta p_L$$

Two types of dynamic instability (Type II) can be distinguished as slow and fast. In both cases assume $\Delta p_L = 0$.

Slow Instability. Theoretically it should be possible to eliminate Δx_F using singular perturbation theory and obtain the linearized slow system as $\Delta \dot{x}_S = A_S \Delta x_S$. The time scale of the phenomenon is so large that the linearized results may not reflect the true picture. For such long time phenomena, a nonlinear simulation is recommended.

Fast Instability. First, we rearrange the variables $\left[\Delta I_g, \Delta \hat{V}_g, \Delta \hat{V}_l \right]$ as $[\Delta I_g, \theta_1, \Delta V_1, \ldots, \Delta V_m | \Delta \theta_2, \Delta \theta_3, \ldots, \Delta \theta_n, \Delta V_{m+1}, \ldots, \Delta V_n] = [\Delta z, \Delta v]$. Next, we assume x_S as constant and load parameters as constant which implies $\Delta p_L = 0$, to get

$$
\begin{bmatrix} \Delta \dot{x}_F \\ 0 \\ 0 \end{bmatrix} = \begin{bmatrix} A_1 & A_2 & A_3 \\ B_1 & B_2 & B_3 \\ C_1 & C_2 & C_3 \end{bmatrix} \begin{bmatrix} \Delta x_F \\ \Delta z \\ \Delta v \end{bmatrix} + \begin{bmatrix} 0 \\ \Delta S_1 \\ \Delta S_2 \end{bmatrix} + \begin{bmatrix} E \\ 0 \\ 0 \end{bmatrix} \Delta u
$$

4.3.5 Analysis Techniques for Dynamic Voltage Stability Studies

It is only recently that the effects of system and load dynamics have begun to be investigated in the context of voltage collapse. The dynamics considered are:

1. Machine and excitation system dynamics including power system stabilizer (PSS)
2. Load dynamics
3. Dynamics of SVC controls and FACTS devices
4. Tap-changer dynamics
5. Dynamics due to load frequency control, AGC, and so on.

While 1, 2, and 3 involve fast dynamics, 4 and 5 represent the slow dynamics. A nice classification process of dynamic voltage stability vis-à-vis static stability is shown in Figure 4.3, where *oad* implies demand; u represents the set points of LFC, AGC and voltage/var controls at substations; x_S represents the slow variables such as the state variables belonging to tap-changing transformers, AGC loop and center of angle variables in the case of a multi-area representation; and x_F represents the fast variables including PSS and governor, induction motor load dynamics, SVC dynamics, and so on, belonging to the generating unit. The overall mathematical model takes the form:

$$
\dot{x}_S = g_s\left(\hat{V}, u \right)
$$
$$
\dot{x}_F = g_F(x_S, x_F, \hat{V}, y, I_g, u, p_L)
$$
$$
0 = h_F(x_S, x_F, \hat{V}, I_g, u, p_L)
$$
$$
y = h_N\left(x_S, x_F, \hat{V}, u \right)
$$

Ignoring the slower AGC dynamics and the faster network transients (60 Hz), we can categorize the variables such that

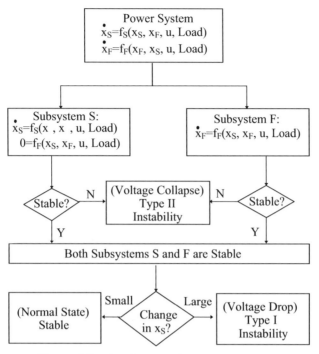

Figure 4.3. Classification of voltage instabilities.

$$x_S = [n_i] \qquad i = 1, \ldots, p$$

where

n_i = transformer tap ratio

$$x_F = \begin{bmatrix} \delta_i \\ \omega_i \\ E'_{qi} \\ E'_{di} \\ E'_{fdi} \\ V_{Ri} \\ R_{fi} \end{bmatrix}$$
rotor angle of ith machine

rotor velocity of ith machine

rotor electrical variables $\quad i = 1, \ldots, m$

excitation system variables

$\hat{V}_i = \begin{bmatrix} V_{Di} \\ V_{Qi} \end{bmatrix}$ Rectangular variables of ith bus voltage or $\begin{bmatrix} \theta_i \\ V_i \end{bmatrix}$

$y = \begin{bmatrix} P_{Gi} + P_{Li} \\ Q_{Gi} + Q_{Li} \end{bmatrix}$ Injected real and reactive power $i = 1, 2, \ldots, n$

$I_{gi} = \begin{bmatrix} I_{di} \\ I_{gi} \end{bmatrix}$ Machine terminal currents in machine reference frame $i = 1, \ldots, m$

$$u = \begin{bmatrix} p_{Mi} \\ V_{ref,i} \\ V_{oi} \end{bmatrix} \begin{array}{l} \text{desired real power of } i^{th} \text{ generator} \\ \text{desired voltage at } i^{th} \text{ generator bus} \\ \text{desired voltage at the bus controlled by tap-changer } i \end{array} \quad \begin{array}{l} i = 1, 2, \ldots, m \\ i = 1, 2, \ldots, p \end{array}$$

p_L = Vector of load parameters

The state variables of the static VAr system (SVC) control and induction motor will appear in x_F if included in the overall model. The example below gives the equations for an m machine n bus system having p tap-changing transformers. Only the synchronous machine and tap-changer dynamics are included.

4.4 VOLTAGE STABILITY ASSESSMENT TECHNIQUES

Now, we discuss the selected voltage stability indices, the VIPI method, the method based on singular value decomposition, condition number of the Jacobian matrix, and the method based on the energy margin. Figure 4.4 depicts the techniques.

VIPI Method. VIPI was developed by Tamura et al. [12] based on the concept of multiple load flow solutions. According to Reference 12, a pair of load flow solutions x_1 and x_2 are represented by two vectors a and b as follows:

$$x_1 = a + b$$

$$x_2 = a - b$$

and are equivalent to:

$$a = \frac{x_1 + x_2}{2}$$

$$b = \frac{x_1 - x_2}{2}$$

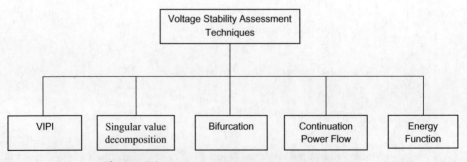

Figure 4.4. Voltage stability assessment techniques.

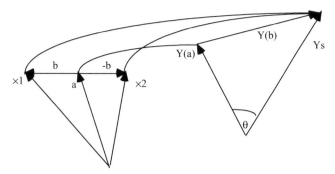

Figure 4.5. Concept of VIPI in the node specification space.

where x_1 is the normal (high) power flow solution, x_2 its corresponding low voltage power flow solution, a is a singular vector in the space of node voltages, and b is a margin vector in the same space.

Define two other vectors, Y_s, and $Y(a)$, called singular vectors in the space of node specifications. The relationship between these vectors is shown in Figure 4.5.

VIPI is defined by the following equation:

$$VIPI = \cos^{-1}\left(\frac{Y_s^T \times Y(a)}{\|Y_s\| \times \|Y(a)\|}\right)$$

where vector Y_s consists of bus injections computed with respect to x_1, but the injection values corresponding to the reactive powers of PV busses are replaced by the squared values of voltage magnitudes. $Y(a)$ consists of bus injections with respect to vector a: $\|X\|$ is the l_2 norm of vector x.

The computation of VIPI is easy once the relevant low voltage power flow solutions are obtained. Exhaustive and simplified methods of locating low voltage power flow solutions of a power system are presented in References 1 and 5, respectively. Generally speaking, finding all the relevant low voltage solutions is a computational burden for practical-sized systems.

Minimum Singular Value. When an operating state approaches the collapse point, the Jacobian matrix of the power flow equations (J) approaches singularityf. The minimum singular value of the Jacobian matrix expresses the closeness of Jacobian singularity. Singular value decomposition method is used to solve the minimum singular value for static voltage stability analysis.

According to the theory of singular value decomposition, the Jacobian matrix of the power flow can be decomposed as:

$$J = U\sum V^T$$

where $J \in R^{2n \times 2n}$ is the Jacobian matrix of the power flow and $\Sigma = \text{Diag}(\sigma_1, \sigma_2, \ldots, \sigma_n)$ with $\sigma_{max} = \sigma_1 \geq \sigma_2 \geq \ldots \geq \sigma_n = \sigma_{min} \geq 0$.

If matrix J has rank r $(r \leq 2n)$, its singular values are the square roots of the r positive eigenvalues of $A^T A$ (or AA^T), U and V are orthonormal matrices of order $2n$, and their columns contain the eigenvectors of AA^T and $A^T A$ respectively. Thus:

$$A V_i = \sigma_i u_i$$
$$A^T u_i = \sigma_i V_i$$

Next, define

$$E_j = u_j V_j^T$$

This equation can be written as:

$$J = \sigma_1 E_1 + \sigma_2 E_2 + \cdots + \sigma_n E_n$$

If we allow

$$J' = \sigma_1 E_1 + \sigma_2 E_2 + \cdots + \sigma_{n-1} E_{n-1}$$

Concerning the l_2 norm of the J matrix, J' is a matrix of rank $n-1$ nearest to the J matrix of rank n, that is, the smallest singular value of a matrix is a measure of the distance between matrices J and J'. The minimum singular value for the power flow equations expresses the proximity of the Jacobian matrix to singularity and can be used as an index for static voltage stability.

Condition Number of the Jacobian Matrix.

The condition number is used in numerical analysis to analyze the propagation of errors in matrix A or vector b in solving variable vector x for the linear equation $Ax = b$. If matrix A is ill-conditioned, even very small variations in vector b (or A) may result in significant changes in solution vector x.

For the linearized load flow equations, the condition number of the Jacobian matrix can be used to measure its conditioning and whether any small variations in vector b (or A) may result in significant changes in solution vector x. For the linearized load flow equations, the condition number of the Jacobian matrix can be used to measure its conditioning and whether any small variations in loads may lead to large changes in bus voltages. If the condition number is greater than a specified threshold, the current operating state is close to the collapse point.

A precise measure of the sensitivity of a linear system solution with respect to matrix A or vector b can be defined as:

$$\text{Cond}_2 (J) = \frac{(\text{max stretch of } A^T A)^{\frac{1}{2}}}{(\text{min stretch of } A^T A)^{\frac{1}{2}}}$$

If matrix A is symmetric with eigenvalues σ_1, σ_2, . . . , σ_n, then Cond_2 (A) is expressed as:

$$\text{Cond}_2\left(\text{A}\right) = \frac{\max|\sigma_i|}{\min|\sigma_i|}$$

For power flow Jacobian matrix J, the value of Cond_2 (J) can give an indication of the condition of J with respect to inversion. A small value of Cond_2 (J) (1~10) refers to a well-conditioned Jacobian matrix (relatively large voltage stability margin); a large value of Cond_2 (J) (>100) means that the operating state is very close to the point of Jacobian singularity and has a low voltage stability margin. The extreme condition is that J is singular and Cond_2 (J) is infinite. Hence, the condition number Cond_2 (J) can be used to measure the proximity of the operating states to voltage collapse.

4.5   VOLTAGE STABILITY INDEXING

Many indices characterizing the proximity of an operating state to the collapse point have been developed. The degeneracy of the load flow Jacobian matrix is used as an index of power system steady-state stability. Under certain conditions, a change in the sign of the determinant of the Jacobian matrix during continuous variations of parameters means that a real eigenvalue of the linearized swing equations crosses the imaginary axis into the right half of the complex plane and stability is lost. Some authors have considered that a change in the sign of the Jacobian matrix probably does not indicate the loss of steady-state stability when there are an even number of eigenvalues whose real parts cross the imaginary axis. Voltage stability is related to multiple load flow solutions.

A proximity indicator for VCPI was defined for a bus, an area, or the complete system as a vector of ratios of the incremental generated reactive power at a generator to a given reactive load demand increase. A different indicator (L index) is calculated from normal load flow results with reasonable computations. The minimum singular value of the Jacobian matrix was proposed as a voltage security index, since the magnitude of the minimum singular value coincides with the degree of Jacobian ill-conditioning and the proximity to collapse point. Based on a similar concept, the condition number of the Jacobian was applied as an alternative voltage instability indicator.

Bifurcation theory was used to analyze static stability and voltage collapse. Static bifurcation of power flow equations was associated with either divergent-type instability or loss of causality. The necessary and sufficient conditions for steady-state stability are based on the concept of feasibility regions of power flow maps and feasibility margins but with high computational efforts. A security measure was derived to indicate system vulnerability to voltage collapse using an energy function for system models that include voltage variation and reactive loads. The conclusion is that the key to applications of the energy method is to find the appropriate Type-1 low voltage solutions.

In addition to the above methods for direct computation of a stability index, some indirect approaches, based on either the continuation methodc or optimization methods, have been developed to compute the exact point of collapse. In applying the continuation methods, different assumptions about load changing patterns are needed.

To sum up, the methods for static voltage instability analysis are based on multiple load flow solutions (VIPI, energy method), load flow results (L index, VCPI), or eigenvalues of the Jacobian matrix (minimum singular value and condition number). While studies on dynamic voltage collapse explain control strategy design (offline applications), static voltage stability analysis can provide operators with guideline information on the proximity of the operating state to the collapse point (online applications). In this case, an index, which can give advance warning about the proximity to collapse point, is useful.

The differential change in voltage at each bus for a given differential change in system load is available from the tangent vector. Therefore, one way to identify the weakest bus with respect to voltage stability limits is to find the bus with the largest dV_i/dP_{total}, where dP_{total} is the differential change in active load for the entire system and is given by

$$dP_{total} = \sum_n dP_{L_i} = \left(S_{\Delta BASE} \sum_n K_{L_i} \cos(\psi_i) \right) d\lambda = Cd\lambda$$

The weakest bus j is therefore given by

$$\left| \frac{dV_j}{Cd\lambda} \right| = \max \left[\left| \frac{dV_1}{Cd\lambda} \right|, \left| \frac{dV_2}{Cd\lambda} \right|, \ldots, \left| \frac{dV_n}{Cd\lambda} \right| \right]$$

When the weakest bus j reaches its steady-state, the voltage stability limit $d\lambda$ approaches zero and the ratio $|dV_j/Cd\lambda|$ will become infinite or, equivalently, the ratio $|Cd\lambda/dV_j|$ tends to zero. The latter ratio which is easier to handle numerically and makes a good voltage stability index for the entire system. Instead of dP_{total}, dQ_{total} can also be used.

The International Council on Large Electric Systems (CIGRE) task force categorizes indices as given operating state-based and large deviation-based. Large deviation indices account for discontinuities such as generator current limiting. Based on the system's ability to withstand load or power transfer increases, an MW or MVAr distance or margin from the operating point to the maximum power transfer point is determined.

Given operating state indices are based on a solved power flow case or an actual real system operating point. Observation of reactive power reserve is a simple and valuable index that can be calibrated to system security.

Determining the index for DC systems is somewhat different. The index may be calculated as $\Delta V = \sum_{i=1}^n V_i - V_{ref}$ where V_i is the voltage at bus i, V_{ref} is the reference bus voltage, and choosing a threshold is the security.

The modal or eigenvalue analysis method is akin to sensitivity analysis but the modal separation provides additional insight. The system partitioned matrix equations of the Newton-Raphson method can be rewritten as

$$\begin{bmatrix} \Delta P \\ \Delta Q \end{bmatrix} = \begin{bmatrix} J_{P_\theta} & J_{PV} \\ J_{Q_\theta} & J_{Q_\theta} \end{bmatrix} \begin{bmatrix} \Delta \theta \\ \Delta V \end{bmatrix}$$

Where the partitioned Jacobian matrix reflects a solved power flow condition and includes enhanced device modeling, by letting $\Delta P = 0$, we can write

$$\Delta Q = \left[J_{QV} - J_{Q_\theta} J_{P_\theta}^{-1} J_{PV} \right] \Delta V = J_R \Delta V$$

where J_R is a reduced Jacobian matrix of the system. J_R directly relates the bus voltage magnitude and bus reactive power injection.

Let λ_i be the ith eigenvalue of J_R and ξ_i and η_i be the corresponding right-column and left-row eigenvectors respectively. The ith modal reactive power variation is

$$\Delta Q_{mi} = K_i \xi_i$$

where

$$K_i^2 \sum_j \xi_{ji}^2 = 1$$

with ξ_{ji} the jth element of ξ_i. The corresponding i^{th} modal voltage variation is

$$\Delta V_{mi} = \frac{1}{\lambda_i} \Delta Q_{mi}$$

The magnitude of each eigenvalue λ_i determines the weakness of the corresponding modal voltage. The smaller the magnitude of λ_i, the weaker the corresponding modal voltage. If $\lambda_i = 0$, the ith modal voltage will collapse because any change in that modal power will cause infinite modal voltage variation.

If all eigenvalues are positive, the system is considered voltage-stable. This is a different dynamic system where eigenvalues with negative real parts are stable. The relationship between system voltage stability and eigenvalues of the J_R matrix is best understood by relating the eigenvalues with the Q-V sensitivity of each bus. J_R can be taken as a symmetric matrix and therefore the eigenvalues of J_R are close to being purely real. If all the eigenvalues are positive, J_R is positive definite and the V-Q sensitivities are also positive, indicating that the system is voltage-stable.

The system is considered voltage-unstable if at least one of the eigenvalues is positive. A zero eigenvalue of J_R means that the system is on the verge of voltage instability. Furthermore, small eigenvalues of J_R determine the proximity of the system to be voltage-unstable.

The participation factor of bus k to mode i is defined as

$$P_{Ki} = \xi_{ki} \eta_{ik}$$

For all of the small eigenvalues, bus participation factors determine the areas close to voltage instability. In addition to the bus participations, modal analysis also calculates branch and generator participations. Branch participations indicate which branches are important in the stability of a given mode. This provides insight into possible remedial actions as well as contingencies which may result in loss of voltage stability. Generator participations show which machines must retain reactive reserves to ensure stability of a given mode.

For a system with several thousand busses it is impractical and unnecessary to calculate all of the eigenvalues. Calculating only the minimum eigenvalue of J_R is not sufficient because there is usually more than one weak mode associated with different parts of the system, and the mode associated with the minimum eigenvalue may not be the most troublesome mode in the stressed system. The m smallest eigenvalues of J_R are the m least stable modes of the system. If the biggest of the m eigenvalues, say mode m, is a strong enough mode, the modes that are not computed can be neglected because they are known to be stronger than mode m. An implicit inverse lopsided simultaneous iteration technique is used to compute the m smallest eigenvalues of J_R and the associated right and left eigenvectors.

Similar to sensitivity analysis, modal analysis is only valid for the linearized model. Modal analysis can, for example, be applied at points along P-V curves or at points in time in a dynamic simulation.

4.6   ANALYSIS TECHNIQUES FOR STEADY-STATE VOLTAGE STABILITY STUDIES

In their early stages, voltage collapse studies mainly concerned steady-state voltage behavior. Voltage collapse is often described as a problem that results when a transfer limit is exceeded. The transfer limit of an electrical power network is the maximal real or reactive power delivered by the system from generation sources to the load areas. Specifically, the transfer limit is the maximal amount of power that corresponds to at least one power flow solution. From the well-known PV or QV curves, one can observe that the voltage gradually decreases as the power transfer amount is increased. Beyond the maximum power transfer limit, the power flow solution does not exist, which implies that the system has lost its steady-state equilibrium point. From an analytical view, the criteria for detecting the point of voltage collapse is the point where the Jacobian matrix of power flow equations becomes singular.

The steady-state operation of the power system network is represented by power flow equations given by

$$F(\theta, V, \lambda) = 0$$

where θ represents the vector of bus voltage angles and V represents the vector of bus voltage magnitudes. λ is a parameter of interest we wish to vary. In general, the dimension of F will be $2n_1 + n_2$, where n_1 and n_2 are the number of PQ and PV busses, respectively.

It should be emphasized that the singularity of the power flow Jacobian $\partial F/\partial x$ is a necessary but not sufficient condition to indicate voltage instability. The method proposed to observe the voltage instability phenomenon is closely related to multiple power flow solutions caused by the nonlinearity of power flow solutions. The drawback of the method is that it relies on the Newton-Raphson method of power flow analysis, which is unreliable in the vicinity of the voltage stability limit.

4.6.1   Direct Methods for Detecting Voltage Collapse Points

This approach tries to find the maximum allowable variation of λ, that is, an operating point (x^*, λ^*) of $F(x, \lambda) = 0$ such that the Jacobian at this point is singular. It solves the following system of equations:

$$G(y) = \begin{bmatrix} F(x,\lambda) \\ F_x^T(x,\lambda)h \\ h_k - 1 \end{bmatrix} = 0$$

This procedure basically augments the original set of power flow equations $F(x, \lambda) = 0$ by $F_x(x, \lambda)h = 0$ where λ is an n-vector with $h_k = 1$. The disadvantages of this approach are:

- The dimension of the nonlinear set of equations to be solved is twice that for the conventional power flow
- The approach requires a good estimate of the vector λ

The advantage is that convergence of the direct method is very fast if the initial operating point is close to the turning point. The enlarged system is solved in such a way that it requires the solution of four $n \times n$ (n is the dimension of the Jacobian $F_x(x, \lambda)$) linear systems, each with the same matrix.

4.6.2   Indirect Methods (Continuation Methods)

Assuming that the first solution (x_0, λ_0) of $F(x, \lambda) = 0$ is available, the continuation problem is to calculate further solutions, (x_1, λ_1), (x_2, λ_2), until a target point is reached, say at $\lambda = \lambda^*$. The ith continuation step starts from an approximation of (x_i, λ_i) and attempts to calculate a next solution. However there is an intermediate step. With predictor, corrector type continuation, the step $i \rightarrow i + 1$ is split into two parts. The first part tries to predict a solution, and the second part tries to make this predicted part converge to the required solution:

Predictor $(x_i, \lambda_i) \rightarrow (\overline{x}_{i+1}, \overline{\lambda}_{i+1})$
Corrector $(\overline{x}_{i+1}, \overline{\lambda}_{i+1}) \rightarrow (x_{i+1}, \lambda_{i+1})$

Continuation methods differ in choice of predictor, type of the parameterization strategy, type of corrector method, and step length control.

Parameterization. The branch consisting of solutions forming a curve in the (x, λ) space has to be parameterized. A parameterization is a mathematical way of identifying each solution on a branch. There are many different kinds of parameterization. For instance, looking at a *PV* curve shows that the voltage is continually decreasing as the load nears maximum. Thus, the voltage magnitude at some particular bus could be changed by small amounts and the solution is found for each given value of the voltage. Here the load parameter would be free to take on any value it needed to satisfy the equations. This is called local parameterization. In local parameterization the original set of equations is augmented by one equation that specifies the value of one of the state variables. In equation form this is expressed as:

$$\begin{bmatrix} F(y) \\ y_k - \eta \end{bmatrix} = 0, \qquad\qquad y = \begin{bmatrix} \theta \\ V \\ \lambda \end{bmatrix}$$

where η is an appropriate value for the kth element of y. Once a suitable index k and the value of η are chosen, a slightly modified N-R power flow method (altered only in that one additional equation and one additional state variable are involved) can be used to solve the set of equations. This provides the corrector needed to modify the predicted solution discussed in the previous section.

A simple example to explain the static analysis procedure follows.

4.7   APPLICATION AND IMPLEMENTATION PLAN OF VOLTAGE STABILITY

Consider the power flow equation defined in the equation below. The vector function F consists of n scalar equations defining a curve in the $n + 1$ dimensional (x, λ) space. Continuation means tracing of this curve. For a convenient graphical representation of the previous solution (x, λ) we need a one-dimensional measure of x. The frequently used measures are:

(i) $|x| = \sum_{i=1}^{n} x_i^2$ (square of the Euclidean norm)
(ii) $|x| = \max_{i=1,n} |x_i|$ (maximum norm)
(iii) $|x| = x_k$ For some index k, $1 \le k \le n$

Power systems generally use the measure of (iii). As Figure 4.6 illustrates, we have a type of a critical solution for $\lambda = \lambda^*$, where for $\lambda > \lambda^*$ there are no solutions. For $\lambda < \lambda^*$ we have two solutions (high voltage solution, low voltage solution). When λ approaches $\lambda^*(\lambda < \lambda^*)$, both solutions merge. At this point, the Jacobian of the power flow solution is singular. In the mathematical literature these points are called turning points, or fold points. An algebraic feature of the turning point given by F_x is

• $F_x(x^*, \lambda^*)$ is singular for rank $<n$
• $F_x(x^*, \lambda^*)/F_\lambda(x^*, \lambda^*)$ has a full rank n and satisfies some nondegeneracy conditions

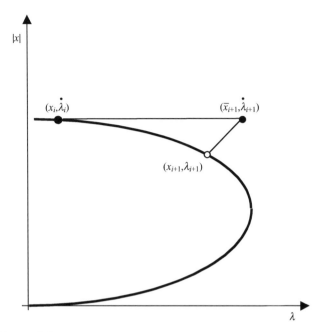

Figure 4.6. The fold type curve including predictor-corrector step.

Several techniques proposed to calculate these points are based on two approaches: direct and indirect.

4.8   OPTIMIZING STABILITY CONSTRAINT THROUGH PREVENTIVE CONTROL OF VOLTAGE STABILITY

The controllable reactive power sources include generators, shunt reactors, shunt capacitors, and on load tap changers of transformers (OLTC). Generators can generate or absorb reactive power depending on the excitation. When overexcited they supply the reactive power, and when under-excited they absorb reactive power. The automatic voltage regulators of generators can continually adjust the excitation. Reactors shunt capacitors and OLTCs are traditionally switched on/off through circuit breakers on command from the operator. Since the early 1980s, advances in flexible AC transmission systems (FACTS) controllers in power systems have led to their application to improve voltage profiles of power networks. Figure 4.7 depicts the control scheme.

The most frequently used devices are Reactive Power Controllers (RPC) and Static Var Compensators (SVC). The RPC connects or disconnects capacitor stages automatically by detecting the phase divergence between the fundamentals of current and voltage. The measured divergence is compared with several segmental set phase divergence regions, and the capacitor contactors will be switched on or off accordingly. SVC, more advanced electronics, provides continuous capacitive and inductive reactive

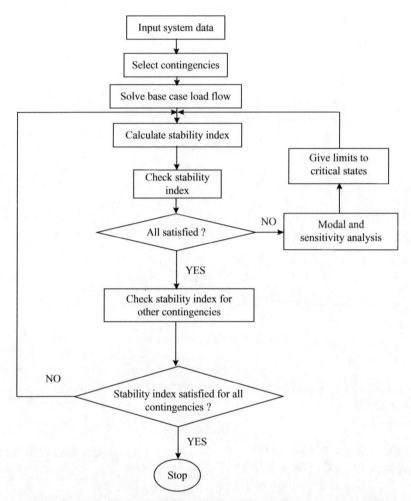

Figure 4.7. Voltage instability preventive control scheme for the power grid.

supply to the power system. The SVC typically consists of a thyristor controlled reactor (TCR), a thyristor switched capacitor (TSC) and AC filters (ACF). From the view system operation, the SVC is equivalent to a controllable reactor and a fixed capacitor. Its output can vary depending on the level of generation and absorption of reactive power needed to maintain its terminal voltage at a certain level. Both the RPC and SVS require significant investments. Since they work only in a local area, the reactive power sources of a network must be coordinated to achieve network voltage stability. A new optimal power flow that includes dynamics of DSOPF is needed.

In an operations planning environment, it is desirable to determine whether a set of dynamic disturbance scenarios is stable or unstable and to assign a measure to its degree of stability, in order to compare the various configurations and determine the transfer level that exhausts the available stability margin. Although transient and voltage stability—due to distinct mechanisms and separate time frames—have sometimes been perceived as decoupled phenomena, in reality they ought to be treated as aspects of coupled short-term power system dynamics [21]. Various researchers have used energy function methods to quickly assess whether a transient disturbance will result in loss of stability. A focus on transient stability assessment methods that yield results are fast but operations planners synthesize the output of simulations resulting in a multitude of time-domain curves into a single, easy-to-interpret number, namely, the energy margin (EM).

For the transmission and distribution system to become more affordable, reliable and sustainable the grid needs to become smarter. During the past few years a considerable number of activities have been carried out in to achieve a smart power grid. Smart grid is envisioned to use all present technologies of voltage stability in transforming the grid intelligently and creating better situational awareness and operation friendliness. At the same time blackouts in the past and reliability of the grid are major issues for system engineers. Features of the DSOPF that aid the development of new OPF with dynamic OPF as constraint, solving performance index, new optimizer DSOPF that incorporate the performance in real time under different control situation are modeled as follows:

1. Obtain load and generation control profile and real time data contribution.
2. Develop and optimize the minimum imbalance between load and generation considering the load generation randomness.
3. Design an optimizer that formulates the OPF tool as a constant where generation availability is a constraint in random space.
4. Design computational intelligence or a combination of classical optimization to solve the scheme problem.
5. Utilize the priority list scheme from dynamic programming concept to rank order load generation balance index.

4.9 ANGLE STABILITY ASSESSMENT

Transient stability studies of a typical 500-bus, 100-machine system require an hour of run time, even for a single contingency. Therefore, direct methods of stability assessment, such as those based on Lyapunov or energy functions, are attractive alternatives. It should be noted that a transient stability study is often more than an investigation of whether the synchronous generators, following the occurrence of disturbance, will remain in synchronism, for it can be a general-purpose transient analysis to investigate the quality of the dynamic system behavior. The transient period of primary interest is the electromechanical transient, usually lasting up to a few seconds in duration. If growing oscillations or the behavior of special controls are a concern, the study may involve a longer transient period. The problem solution uses one of the family of solutions depicted in Figure 4.8.

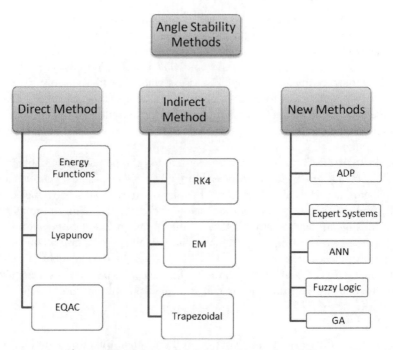

Figure 4.8. Overview of methods for angle stability.

In actual system operation, the parameters and loading conditions are quite different from the assumptions made in the planning stage. As a result, to ensure power system security against possible abnormal conditions due to contingencies, the system operator needs to simulate contingencies in advance, assess the results, and take preventive control action if required. The equation for solving this can be stated and modeled as:

$$\mathrm{M}d^2\delta/dt^2 = \mathrm{P_m} - \mathrm{P_e}$$

which is written in state space as

$$d\delta/dt = f(\delta, u)$$

$$d^2\delta/dt^2 = \text{acceleration}$$

where δ = angle

u = Control (power system generator taps, and so on)

$\mathrm{P_m}$, $\mathrm{P_e}$ = mechanical and electrical power

To solve the time domain stability problem, numerical methods are utilized based on predictive and corrective methods and to provide suggestions for further interpretation, including:

1. Numerical method: RK4, Trapezoidal method, modified Euler method, Theta method. The Euler approach converts differential equations from digital to analogue; numerical methods solve numerical method technique by analyzing decoupled first- and second-order state differential equation with the predictor corrector technique; the trapezoidal method is faster for still networks than other methods. All of these technique have similarities: the initial condition of the network, step-size selection, and integration application.

Although these methods are used today, they lack characteristics such as the robustness, scalability, stochasticity, predictivity, and adaptability of the future grid, which is completely absent in the current power system grid network.

2. Artificial Intelligence Method. This technique uses heuristic and computational intelligence to estimate the strability margin and compuatational indices. The method includes the expert system for classifying contingencies, identifying and estimating the instability margin, and so on.

The neural network is also used as a classifer for different stability margins based on network features and parameters contributing to the estimation of instabilities. Other AI methods include particle swarm optimization, genetic algorithm. These methods are used to model the inherent and stochastic elements in the network. Several of these techniques have been tested in the literature and they form the basis for a new methodology for angle stability margin or assessment in real time.

An important techique for stability assessment includes the dynamic security assessment used over several years. These method is useful as it can take in real time data measurement, unlike other methods.

3. Evaluation of dynamic security assessment and its potential for real-time stability in a smart grid environment. DSA includes the study of power system oscillation in terms of generating power plants and network configuration under different contingencies. To classify the impact of the contingency, an assessment of the energy margin of unit impact on loss of line or unit consists of measures of the potential and kinetic energy associated with the machines in the system with respect to a reference machine, or COA. The sum of the kinetic and potential energy is compiled and compressed to a given threshold in order to determine the network impacts, and rank and classify them.

4.9.1 Transient Stability

Consider an autonomous system described by the ordinary differential equation

$$\dot{x} = F(x)$$

where $x = x(t)$, and $F(x)$ are n-vectors. $F(x)$ is generally a nonlinear function of x. Stability in the sense of Lyapunov is referred to as an equilibrium state. The equilibrium state is defined as the state x_e at which $x(t)$ remains unchanged for all t. That is,

$$\dot{x}_e = F(x_e) = 0$$

The solution for x_e is a fixed state since $F(x)$ is not an explicit function of t. For convenience, any nonzero x_e is to be translated to the origin ($x = 0$.) That is, replace x by $x + x_e$ to have

$$\frac{d}{dt}(x + x_e) = F(x + x_e) = f(x)$$

which gives

$$\dot{x} = F(x)$$

Note that the current x differs from the old one by x_e. As can be seen later from the definitions, this translation does not affect the stability of the system. Thus, the origin is always an equilibrium state. Note that t may be any independent variable including the time.

Let $a = d^2\delta/dt^2$ denote the acceleration of the system with $a < 0$ above the P_m line and $a > 0$ below it after fault clearance. There are two angles: $\delta_s = \sin^{-1}P_m/A(X_c)$ and $\delta_u = \pi - \delta_s$ that correspond to zero acceleration. δ increases from δ_0 with $\omega = 0$ due to decreasing acceleration, until reaching $\delta_m < \delta_u$, at which point the speed, ω, is zero. Since the acceleration is negative at δ_m, δ begins to decrease until the machine speed is zero at an angle less than δ_s and then comes back because $a > 0$. As such, the power angle swings back and forth around δ_s. This is the case for $E - V < 0$ at δ_u because δ_u is unreachable (ω is imaginary).

However, one fault clearance may result in large A_1 which makes δ cross δ_u with $\omega \geq 0$. Then, δ increases further without return due to $a > 0$. This is the case for $E - V \geq 0$. Thus, the transient is stable if the power angle swings around δ_s and unstable otherwise. This definition makes it possible to assess the stability with other techniques such as transient energy function (TEF) as follows:

(a) The transient is stable if $E - V < 0$ at δ_u, and large magnitude yields better stability
(b) The transient is unstable if $E - V \geq 0$ at δ_u

4.9.2 Stability Application to a Practical Power System

Since the application of the direct method to actual power systems is quite difficult, it becomes necessary to make a number of simplifying assumptions. To date, the analysis has been mostly limited to power system representation with generators represented by classical models and loads modeled as constant impedances. Recently several attempts to extend the method now include more detailed load models.

In a multi-machine power system, the energy function $\mathbf{E}$ describing the total system transient energy for the post-disturbance system is given by:

$$\nabla = -\frac{1}{2}\sum_{i=1}^{n} J_i \omega_i^2 - \sum_{i=1}^{n} P'_{mi}(\theta_i - \theta_i^s) -$$

$$\sum_{i=1}^{n-1}\sum_{j=i+1}^{n}\left[C_{ij}(\cos\theta_{ij} - \cos\theta_{ij}^s) - \int_{\theta_i^s+\theta_j^s}^{\theta_i+\theta_j} D_{ij}\cos\theta_{ij}d(\theta_i + \theta_j) \right]$$

where

θ_i^s = angle of bus i at the post-disturbance SEP

$J_i = 2H_i\omega_0$ = per unit moment of inertia of the ith generator

The transient energy function consists of the following four terms:

1. $\frac{1}{2}\sum J_i \omega_i^2$: change in rotor kinetic energy of all generators in the COI reference frame

2. $\sum P'_{mi}(\theta_i - \theta_i^s)$: change in rotor potential energy of all generators relative to COI

3. $\sum\sum C_{ij}(\cos\theta_{ij} - \cos\theta_{ij}^s)$: change in stored magnetic energy of all branches

4. $\sum\sum \int D_{ij}\cos\theta_{ij}d(\theta_i + \theta_j)$: change in dissipated energy of all branches

The first term is called the kinetic energy (E_{ke}) and is a function of generator speeds only. The sum of terms 2, 3, and 4 is called the potential energy (E_{pe}) and is a function of generation angles only.

The assessment procedure has three steps:

Step 1: Calculate the critical energy V_{cr}

Step 2: Calculate the total system energy at the instant of fault-clearing E_{cl}

Step 3: Calculate the stability index: $V_{cr} - V_{cl}$; the system is stable if the stability index is positive

A time domain simulation is run up to the instant of fault clearing to obtain the angles and speeds of all generators. These are used to calculate the total system energy (V_{cl}) at fault clearing. The flowchart of the TEF for transient stability analysis is shown in Figure 4.9.

4.9.3 Boundary of the Region of Stability

The calculation of the boundary of the region of stability, V_{cr}, is the most difficult step in applying the TEF method. Three different approaches are briefly described here.

Algorithm for the closest unstable equilibrium point (UEP) approach.
Early papers on the application of the TEF method for transient stability analysis used the following approach to determine the smallest E_{cr}.

Step 1: Determine all of the UEPs by solving the post-disturbance system steady-state equations with different initial values of bus angles

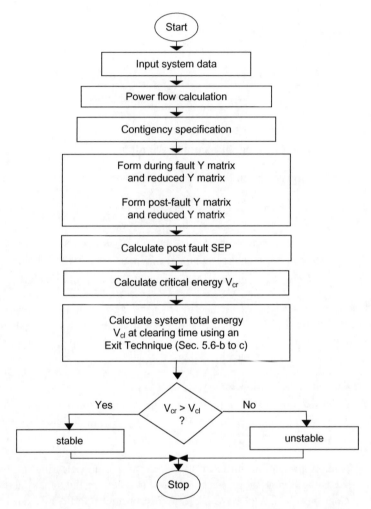

Figure 4.9. Flowchart of TEF for transient stability analysis.

Step 2: Calculate system potential energy at each of the UEPs obtained in step 1; the critical energy V_{cr} is given by the system at the UEP, which results in the minimum potential energy

Because this approach computes the critical energy by implicitly assuming the worst fault location, the results are very conservative.

Algorithm for the controlling UEP approach. The degree of conservatism introduced by the closest UEP approach is such that the results are usually of little practical value. The controlling UEP approach removes much of this conservatism by computing the critical energy depending on the fault location. This approach is based on the observation that the system trajectories for all critically stable cases get close to the UEPs that are closely related to the boundary of system separation. The UEPs are called the controlling or relevant UEPs.

The essence of the controlling UEP [6,7] method as shown by Chiang uses the constant energy surface through the controlling UEP to approximate the relevant part of the stability boundary (stable manifold of the controlling UEP) to which the fault-on trajectory is heading.

For any fault-on trajectory $x_f(t)$ starting from a point $p \in A(x_s)$ with $V(p) < V(\hat{x})$, if the exit point of the fault-on trajectory lies in the stable manifold of $\hat{x}$, the fault-on trajectory must pass through the connected constant energy surface $\partial V_c(\hat{x})$ before it passes through the stable manifold of $\hat{x}(W^S(\hat{x}))$ (thus exiting the stability boundary $\partial A(\hat{x}_S)$). Therefore, the connected constant energy surface $\partial V_c(\hat{x})$ can be used to approximate the part of the stability boundary $\partial A(\hat{x}_S)$ for the fault-on trajectory $x_f(t)$. The four steps in the computation process are:

1. Determine the controlling UEP, x_{co} for the fault-on trajectory $x_f(t)$
2. Setting the critical energy V_c as the value of the energy function $V(\cdot)$at the controlling UEP, that is, $V_c = V(x_{co})$
3. Calculate the value of the energy function $E(\cdot)$ at the time of fault clearance (say, t_{cl}) using the fault-on trajectory $V_f = V(x_f(t_{cl}))$
4. If $V_f < V_c$, then the post-fault system is stable, otherwise, it is unstable

The key is how to find the controlling UEP for a fault-on trajectory. Much of the recent work on the controlling UEP method is based on heuristics and simulations. A theory-based algorithm to find the controlling UEP for the classical power system model with transfer conductance G_{ij} is discussed in the next section.

The energy function is of the form:

$$
\begin{aligned}
V(\delta,\omega) = \sum_{i=1}^{n-1}\sum_{j=i+1}^{n} & \left\{ \left(\frac{1}{2M_T} M_i M_j (\omega_{in} - \omega_{jn}) \right)^2 \right. \\
& - \frac{1}{M_T}(P_i M_j - P_j M_i)(\delta_{in} - \delta_{jn} - \delta_{in}^S + \delta_{jn}^S) \\
& \left. - V_i V_j B_{ij} \{ \cos(\delta_{in} - \delta_{jn}) - \cos(\delta_{in}^S - \delta_{jn}^S) \} \right\} \\
& - V_i V_j G_{ij} \frac{\delta_{in} + \delta_{jn} - (\delta_{in}^S + \delta_{jn}^S)}{\delta_{in} - \delta_{jn} - (\delta_{in}^S - \delta_{jn}^S)} \{ \sin(\delta_{in} - \delta_{jn}) - \sin(\delta_{in}^S - \delta_{jn}^S) \} \\
= V_p(\delta) & + \frac{1}{2M_T} \sum_{i=1}^{n-1}\sum_{j=1}^{n} M_i M_j (\omega_{in} - \omega_{jn})^2
\end{aligned}
$$

where $M_T = \sum_{i=1}^{n} M_i$, $x^S = (\delta^S, 0)$ is the stable equilibrium point (SEP) under consideration. Note that $V(\delta,\omega)$ represents the energy margin, while V_i and V_j are nodal voltages within the network.

4.9.4 Algorithm to Find the Controlling UEP

The reduced system is

$$\dot{\delta}_{in} = \frac{1}{M_i}(P_i - P_{ei}) - \frac{M_i}{M_n}(P_n - P_{en})$$
$$= f_i(\delta) \qquad i = 1, 2, \cdots, n-1$$

The algorithm for finding the controlling UEP consists of five steps:

Step 1: From the fault-on trajectory $(\delta(t), \omega(t))$, detect the point δ^* at which the projected trajectory $\delta(t)$ reaches the first local maximum of $V_p(.)$, and then compute the point δ^- that is one step ahead of δ^* along $\delta(t)$, and the point δ^+ that is one step after δ^*

Step 2: Use the point δ^* as initial condition and integrate the post-fault reduced system Equation (5–31) to find the first local minimum of $\sum_{i=1}^{n}|f_i(\delta)|$, say at δ_0^*

Step 3: Use δ^- and δ^+ as initial conditions and repeat Step 2 to find the corresponding points, say δ_0^- and δ_0^+, respectively

Step 4: Compare the values of $|f(\delta)|$ at δ_0^-, δ_0^*, and δ_0^+; use the one with the smallest value as the initial guess to solve (5–31), $f_i(\delta) = 0$, say the solution is δ_{co}

Step 5: The controlling UEP for the fault-on trajectory is $(\delta_{co}, 0)$

4.9.5 Process Changes in Design of DSA for the Smart Grid

We now give the possible options for designing DSA for the smart grid.

Option 1. Real Time DSA with PMU is currently being used by some researchers, but we propose a modification that includes system topology in time stamp for a new topology of the power system with or without RER in three steps:

Step 1 Develop a computer index for assessing stability based on different scheduled parameter changes

Step 2 Develop a new ranking procedure in time interval and utilize some form of probability function

Step 3 Identify source of instability and margin of stability for next corrective or preventive action

Option 2. The second option includes variability in the definition of system components and devices as well as the model of load in stochastic form. Doing so places a construct on the network solution as a probabilistic load flow. The model of the generators are RER mixed generation (some deterministic and some probabilistic). The aggregate of each requires some correlative probability distribution function. This dynamic equation is now treated as a probabilistic method with results given in terms of mean and variance. We then apply reversed function computation to address the state of margin calculation computation. The interpretation of the results also accounts for new variability. Figure 4.10 shows the form. The three steps are:

1. Design nominal integration with variation time to complete values of P as function of $\delta(t)$
2. Analyze the algorithm or differential equation coupled
3. Interpret results and display

Option 3. Research is in progress to include the variability of RER and stochastic load with the appropriate probability distribution function to establish a probabilistic equation and couple it with the differential equation. As before, at each time step, corresponding values of power from control generation and RER using computational intelligence or artificial neural network provide prediction and estimation of instability.

4.10 STATE ESTIMATION

Energy management systems (EMS) run in real time to compute and maintain security of operation at minimum cost. The power system measurements provide information to the SE program for processing and analysis. The functions performed include topology processing which gathers data about breakers and disconnect switches. State estimation (SE) of voltage and angles are obtained for all busses using the weighted least square (WLS) method. Detection of inaccurate data is obtained from network parameters, tap changing transformers, shunt capacitors, or breakers. The general SE model discussed below is based on steady-state functions. Many of the algorithms in use contain problems including:

1. Convergence problem: traditional SE may not converge if the power system state changes faster than the SCADA data
2. Possibility of missing critical data from communication channels
3. System parameter data base with many errors
4. Report changes in system parameters or devices such as breaker, taps, or disconnected switches

This is true because when topology changes are undefined, the estimation is likely to fail because the SE develops on network topology, bus admittance matrix which may fail to lead to invertible Y-bus matrix. The explicit modeling of switches facilitates bad data analysis when topology errors are involved (incorrect status of switching devices). In this case, SE is performed on a model in which parts of the network can be represented physically, allowing for the inclusion of measurements made on zero impedance branches and switching devices. The conventional states of bus voltages and angles are augmented with new state variables. Observability analysis is extended to voltages at bus sections and flows in switching devices; if their values can be computed from the

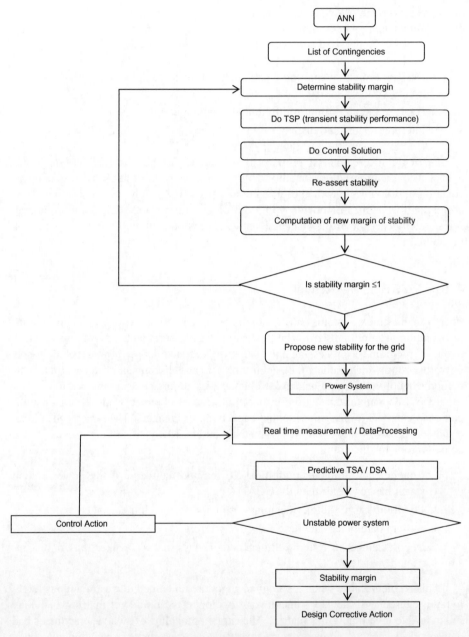

Figure 4.10. DSA for the smart grid.

available measurements, they are considered to be observable. For a zero impedance branch or a closed switch, the following constraints or pseudo- measurements, are included in SE [14]:

$$V_k - V_m = 0 \quad \text{and} \quad \theta_{km} = \theta_k - \theta_m = 0$$

In this case, P_{km} and Q_{km} are used as additional state variables independent of the complex nodal voltages $V_k e^{j\theta_k}$ and $V_k e^{j\theta_m}$, since Ohm's law (in complex form) cannot be used to compute the branch current as a function of these voltages.

For open switches ($z_{km} = \infty$), the same additional state variables are included in SE.

No pseudomeasurements are added in the case of switches with unknown status. Situations occur in which the incorrect status of a switching device can affect SE convergence. In these cases it may be preferable to treat such status as unknown and proceed with SE, which hopefully will include the estimation of the correct status.

The ideas above can be extended to branches with unknown impedances [11]. Branch impedance z_{km} is unknown, whereas for simplicity all branches incident to k and m are assumed to have known impedances. As with zero impedance branches and with closed/open breakers, Ohm's law cannot be used to relate the state variables $V_k e^{j\theta_k}$ and $V_k e^{j\theta_m}$ associated with the terminal nodes k and m with the branch complex power flows $P_{km} + jQ_{km}$, and $P_{mk} + jQ_{mk}$. These power flows can be used as additional states, although they are not independent, since they are linked by the constraint $I_{km} + I_{mk} = 0$ which can be expressed by the two following pseudomeasurements:

$$P_{km}V_m + (P_{mk}\cos\theta_{km} - Q_{mk}\sin\theta_{km})V_k = 0$$
$$Q_{km}V_m + (P_{mk}\sin\theta_{km} + Q_{mk}\cos\theta_{km})V_k = 0$$

A power injection measurement at node k can be expressed as the summation of the flow state variables P_{km}, Q_{km} and the flows in all other branches incident to k. Since only the flows in regular branches are functions of the nodal state variables, the unknown impedance will not form part of the measurement model. A similar analysis holds for power injection measurement at nodal point and power flow measurements made in the unknown impedance branch. Once the network state is estimated, the value of the unknown parameter can be computed from the estimates.

Complex network elements, such as a transmission line π equivalent model, require consideration of other constraints (pseudomeasurements) in addition to the inclusion of flow state variables. Consider, for example, the equivalent π model, where the series branch impedance is to be estimated. In this case, power flows P_{km}, Q_{km}, P_{mk} and Q_{mk} are considered to be additional states. Then, the terminal power flows P_{km}, Q_{km}, P_{mk} and Q_{mk} are expressed in terms of the new state variables rather than as a function of the terminal bus voltages; as a consequence, the series branch impedance will not appear in the measurement model, and it can be written as

$$P_{km} = P_{kk} + P'_{km} \quad \text{and} \quad Q_{km} = Q_{kk} + Q'_{km}$$
$$P_{mk} = P_{mm} + P'_{mk} \quad \text{and} \quad Q_{mk} = Q_{mm} + Q'_{mk}$$

In these equations, note that the power flows P_{kk}, Q_{kk}, P_{mm} and Q_{mm} are written in terms of the shunt parameters as usual. The bus injection measurements at busses k and m are expressed in terms of the terminal power flows as described above. These added states are not entirely independent, so it is necessary to include the following relationship in the model:

$$P'_{km} V_m + (P'_{mk} \cos\theta_{km} - Q'_{mk} \sin\theta_{km}) V_k = 0$$
$$Q'_{km} V_m + (P'_{mk} \sin\theta_{km} + Q'_{mk} \cos\theta_{km}) V_k = 0$$

The constraint linking the state variables in the case of a balanced π model is $y^{th}_{kk} = y^{th}_{mm}$. Expressing the shunt admittances in terms of the corresponding active and reactive power flows yields two pseudomeasurements:

$$P_{kk} V_m^2 - P_{mm} V_k^2 = 0$$
$$Q_{kk} V_m^2 - Q_{mm} V_k^2 = 0$$

The term static state estimation refers to the process of computing solutions of the basic load flow problem using online data telemetered periodically to the energy control center. Data exchanges with neighboring systems to develop an external network equivalent model are facilitated if each system has an online state estimator. An external equivalent representation is required to perform online state contingency analysis. Without an external equivalent model, online SE will be limited to monitoring voltage levels, phase angles, line flows, and network topology.

In load flow, the input/load variables describe the steady-state behavior of the system. In actual online systems it is not usually possible to measure these inputs and loads directly, because loads are net injection quantities, each of which is the sum of several power flow quantities. A single measurement of power flow on a transmission line requires less instrumentation and may be useful in contributing to the load flow solution. In principle, one can measure any meaningful set of system quantities and use those measurements as inputs to a system of equations whose solution is the values of state variables (bus voltage magnitudes and angles). We can compute net bus injections from the estimated state variables.

4.10.1   Mathematical Formulations for Weighted Least Square Estimation

Because of redundant measurements, the solution $\hat{\mathbf{x}}$ of $\mathbf{x}$ is obtained by minimizing the WLS performance index J given by:

$$J = [\mathbf{z} - h(\mathbf{x}, \mathbf{p})]^T \mathbf{R}^{-1} [\mathbf{z} - h(\mathbf{x}, \mathbf{p})]$$

with respect to $\mathbf{x}$. The vector $\mathbf{p}$ is assumed to be known exactly. Hence, we can drop $\mathbf{p}$ from the above expressions. The concise SE problem statement becomes:

Given $\mathbf{z} = h(\mathbf{x}) + \mathbf{v}$ such that $E(\mathbf{v}) = 0$, $E(\mathbf{v}\mathbf{v}^T) = \mathbf{R}$ compute the best estimate $\hat{\mathbf{x}}$ of $\mathbf{x}$ which minimizes:

$$J = [\mathbf{z} - h(\mathbf{x})]^T \, \mathbf{R}^{-1} [\mathbf{z} - h(\mathbf{x})]$$

with respect to $\mathbf{x}$. At the minimum of J we should expect that:

$$\left. \frac{\partial J}{\partial \mathbf{x}} \right|_{\hat{x}} = 0$$

where $\hat{x}$ is the state vector at the minimum of J and is referred to as the best estimate of $\mathbf{x}$. Given the above definition of J, we assert that the zero gradient condition just stated will yield the n-dimensional vector equation:

$$0 = \mathbf{H}^T (\hat{\mathbf{x}}) \mathbf{R}^{-1} (\mathbf{z} - h(\hat{\mathbf{x}}))$$

The necessary conditions for solution are a set of nonlinear algebraic equations requiring an iterative solution method, that is, following the Newton-Raphson method, we linearize the system equations around a nominal value of the state vector $\mathbf{x}$.

Statistical Properties of State Estimator Outputs. The expected value of the optimal estimate vector $\hat{\mathbf{x}}$ is given by:

$$E(\hat{\mathbf{x}}) = \mathbf{x}$$

Moreover, the covariance matrix of $\hat{\mathbf{x}}$ is given by:

$$\mathrm{cov}(\hat{\mathbf{x}}) \left[\mathbf{H}^T \mathbf{R}^{-1} \mathbf{H} \right]^{-1}$$

We are also interested in the expected values of the index (J/m) corresponding to the fit of estimates of measured quantities to the measurements themselves. This can be given by:

$$E(J/m) = \frac{m - n}{n}$$

For $m - n$ (no redundancy), $E(J/m) = 0$ and the estimates fit the measurements perfectly. For $m \to \infty$ (infinite redundancy), $E(J/m) \to 1$ and the estimates approach the true value. In addition, the index J'/m corresponds to the fit of the estimates to the true values of the noisy measurements. This can be given by:

$$E(J'/m) = n/m$$

For $m = n$, we get $E(J'/m) = 1$, and for $m \to \infty$, $E(J'/m) = 0$, $m \to 0$, d.n.e.

In general the index J is chi-square distributed with $m - n$ degrees of freedom.

4.10.2   Detection and Identification of Bad Data

Redundant measurement information allows us to identify and locate bad data consisting of gross measurement errors and/or large modeling errors, for example, wrong network topology, or large parameter errors. Without redundancy the estimates will fit the data perfectly, but it does not provide ways to locate bad data. With redundancy, the WLS algorithm will try to minimize the performance index J (or J_p as the case may be). In the absence of bad data and parameter errors, the expected value of J/m is $(m - n)/m$. Thus we have a ready means to check if the data lies within its postulated error bounds. If $\dfrac{J}{m} \gg \dfrac{m-n}{m}$, then we can be sure that something is wrong. This is the detection step. In this case, we have to look for the source of trouble by means of bad data identification. The four sources of problems are:

1. Gross measurement errors
2. Small modeling errors
3. Small parameter errors
4. Inaccurate knowledge of measurement variances

In a real situation (and especially in initial implementation phases) all of these problems will occur simultaneously. The solution key consists of two primary considerations:

- Exploitation of the structure of power flow equations and associated sensitivities
- Creative hypothesis testing

Extra care is required to make the bad data identification step workable. This is achieved in the two steps, pre- and postestimation analysis, described below.

4.10.3   Pre-Estimation Analysis

Before a given snapshot measurement, vector z of the system is processed in the WLS algorithm. Its components can undergo a series of so-called consistency tests with the following objectives:

- Detection of obviously bad measurements
- Detection of obviously bad network topology
- Classification of data as (a) valid, (b) suspect, and (c) raw
- Tuning the measurement variance values

We discuss some of these topics as follows.

Detection of Obviously Bad Measurements.   In this preliminary stage, measurements whose values are outside reasonable limits are automatically discarded. For example, line flow limits can be set at twice the theoretical capacity of the line. Power factors, voltage levels, and so on can be safely limited. In most cases, almost all of the bad data will be in this category and can be quickly discarded.

Detection of Obviously Bad Network Topology. Normally, a special network configuration program constructs the system network on the basis of breaker status information. Open lines are not represented in the model forwarded to the state estimator, that is, all lines that are in the model used by the estimator should be closed (energized). However, cases may occur where a breaker is closed but has an open disconnect switch. If the disconnect switch status is not reported, the line is mistakenly assumed to be energized. One way to check for this anomaly is to see if the power flow on the line is zero at both ends. If this is the case then the line is most probably open. Other cases of bad topology may be detected from incoming data and are usually peculiar to the system being analyzed.

Classification of Raw Data. Because of the structure of load flow equations a considerable number of hypothesis consistency tests can be conducted to verify the validity of most of the data and to tune the values of the measurement variables. Typical examples of these tests are:

1. Line flow measured at both ends: For real flows, the magnitudes of flows from both ends differ only by the amount of line losses evaluated from:

$$T_{Loss} = T_{ij} + T_{ji}$$
$$= g_{ij}\left(V_i^2 + V_j^2 - 2V_iV_j\cos(\delta_i - \delta_j)\right)$$
$$\approx V_i^2 g_{ij}\left(T_{ij}/b_{ij}\right)^2$$

   As a result, we conclude that $T_{ij} + T_{ji} - T_{Loss} = e_{ij}$ where e_{ij} is an error whose variance is approximately the sum of the variances of the two line flow measurements under consideration. If the ratio

$$\frac{e_{ij}^2}{\sigma_{T_{ij}}^2 + \sigma_{T_{ji}}^2} \leq 9.0,$$

   then the two measurements under consideration are consistent with one another. This assures that the combined errors of both measurements are within the limit. This statement is true with a probability of $\approx 97\%$.

2. Real and reactive line flows measured at both ends, together with a voltage measurement at one end: This information allows computation of the real and reactive power flows at the end and comparison with the measured values of the same quantities; easily validates or invalidates the consistency of the measurements under consideration.

3. Bus injection and line flows measured at same and/or opposite bus ends: Since an injection measurement is equal to the sum of the corresponding line flow measurements, it is a quick consistency check.

4. Local estimators: Dividing the overall network into a set of small observable networks allows rapid computation of SEs for these small networks to check for potential locations of bad data.

5. Pseudo-measurements: By definition, the net power injection at the many unconnected transmission busses is zero and thus is not measured; these zero injections are exact and can be used in the state estimator as pseudo-measurements with very small variances; Alternatively, reformulate an optimization problem to minimize the WLS error subject to equality constraints imposed by the pseudo-measurements; in either case, the pseudo-measurements are automatically valid and useful measurements that are very useful.

We now discuss the process of pre-estimation analysis.

Denote the set of measurements used in a particular consistency test of the problems above by S_i. Let a_i^2 and e_i^2 be the respective overall variance and squared error associated with the test. Let a^2 be a consistency threshold such that if $e_i^2 > a^2$, then the set of measurements in S_i is inconsistent. Otherwise, it is consistent. If the measurements are consistent they are declared valid. If the tests indicate inconsistency, only those measurements in S_i that have not been previously validated are declared suspect. If, by this process S_i contains strictly one suspected measurement, then that measurement is bad and is deleted from the measurement vector.

Figure 4.11 shows the overall block diagram for pre-estimation analysis. At the end of the analysis every measurement will either be raw, valid, suspect, or bad. Only bad measurements are discarded. The variances of valid measurements are slightly modified to reflect information derived from the consistency analysis. Raw measurements are those for which a consistency test cannot be performed, that is, they do not belong to any set S_i. This is usually the case for nonredundant portions of the measurement system. Finally, suspected measurements will contain the desired set of yet-undetermined bad data. The final decision on the suspected measurements is performed in the postestimation analysis process.

4.10.4 Postestimation Analysis

Postestimation analysis looks at the results of state and parameter estimation and attempts to establish hypotheses for the most probable causes of poor performance, if any. This is based on the analysis of the normalized measurement residuals defined as:

$$r_i' = \frac{z_i - h(\hat{\mathbf{x}}, \hat{\mathbf{p}})}{\rho_i}, \quad i = 1, \dots, m$$

where $\rho_i^2 = \text{var}(z_i - h_i(\hat{\mathbf{x}}, \hat{\mathbf{p}}))$.

Obviously ρ_i^2 is the i-th diagonal term of the covariance matrix S defined earlier.

On average, $|r_i'| = 1.0$. Statistically, $|r_i'|$ can vary from zero to three with a high probability. This is true only when all data lie within the specified statistical accuracies. If

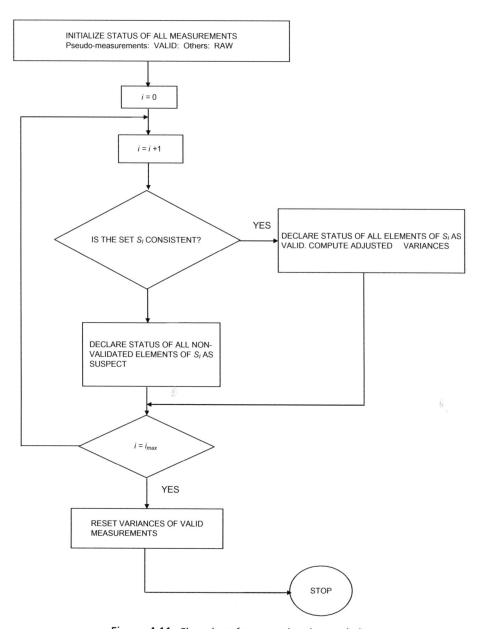

Figure 4.11. Flow chart for pre-estimation analysis.

there are bad data in the measurements and/or parameters, some of the normalized residual terms will have a large magnitude. In many cases, the measurement with the largest normalized residual is a bad measurement, but this has not been mathematically proven. However, we can show that if measurement z_i is bad and if it is a redundant measurement, its normalized residual will be large. In practice, the computation is time-consuming. Therefore, without much loss in information, the residuals defined as: $r_i' = [z_i - h(\hat{x}, \hat{p})]/\sigma_i, i = 1, \ldots, m$ are analyzed instead of the normalized residuals defined above. Usually $r_i^2 \leq 1.0$ is on the average. However, a bad measurement which is redundant will have a large residual.

Without parameter errors and pre-estimation analysis, all measurements are classified as raw with a small probability that some of them are bad. By analyzing the residuals one can start the hypothesis testing process by discarding the measurement with the largest residual and performing the estimation process again. If this fails to yield acceptable estimates, the discarded measurements are put back and the measurement with the largest residual in the new estimate is removed. This process is repeated until a satisfactory answer is obtained. Obviously, successful identification of the bad data will occur only if a single bad measurement is present. For two or more bad measurements this process may fail. In this case, the above process is repeated by discarding two or more measurements with large residuals at a time and so on. However, in this case, computational times will start to become unacceptably excessive.

Hypothesis Testing for the Identification of a Single Bad Data Point. With noisy parameter values, the preceding bad data detection process will be even less attractive because the noisy parameters themselves cause large measurement residuals. Thus, we rely heavily on pre-estimation and bad data analysis. In this case, we assume that bad measurement data resides with high probability among the suspected measurements (see Fig. 4.12). Normally, suspected measurements in a set S_i are highly correlated to one another with minimal or no sensitivity to parameter errors. Thus, we can implement the algorithm whose flowchart is shown in Figure 4.13. In this algorithm, following state and parameter estimation stages, the largest unacceptable residual among suspected measurements in every consistency set S_i points to a bad measurement. Small and acceptable residuals of previously suspected measurements cause the corresponding measurements to be declared valid. This is iterated several times until the system performance reaches acceptable limits.

4.10.5 Robust State Estimation

We consider an approach to SE that automatically compensates for the effects of bad data. The basic concept involves the smearing property of WLS estimation. Recall that in WLS if a measurement is bad the tendency is to spread the effect of the bad measurement residuals over the rest of the system. To develop this concept we write:

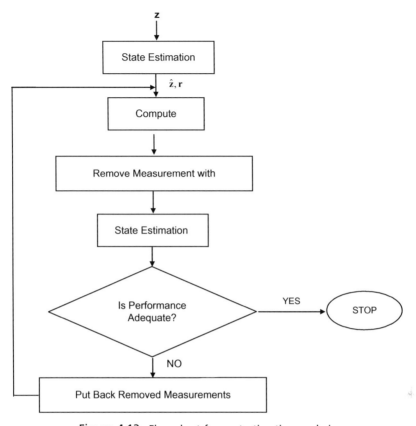

Figure 4.12. Flow chart for postestimation analysis.

$$J = (z - h(\hat{\mathbf{x}}))^T \mathbf{R}^{-1} (z - h(\hat{\mathbf{x}}))$$

$$= \sum_{i=1}^{m} \left(\frac{z_i - h_i(\hat{\mathbf{x}})}{\sigma_i} \right)^2$$

$$= \sum_{i=1}^{m} (r_i)^2$$

$$= \sum_{i=1}^{m} J_i$$

Assume that if r_i for some j is bad, then its real error variance is much larger than the assumed variance. In trying to minimize J, the WLS algorithm attempts to make the terms J_i as equal as possible to each other. In robust estimation, the functions J_i are quadratic in the neighborhood of the origin and flat as the residuals r_i become large. An example is the following J_i:

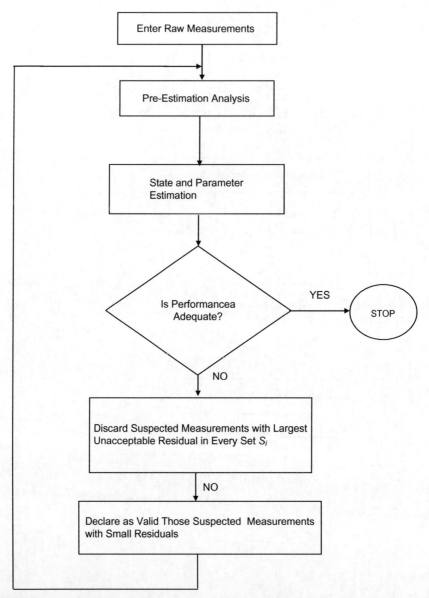

Figure 4.13. Overall flowchart for online, state, and parameter estimation, and bad data analysis.

$$J_i = \begin{cases} r_i^2, & |r_i| \le a \\ 4a^{\frac{3}{2}}|r_i|^{\frac{1}{2}} - 3a^2 & |r_i| \ge a \end{cases}$$

where a is a variable parameter. In this formulation, as r_i becomes large, J_i increases as $|r_i|^{1/2}$. Thus, large residuals will not strongly influence the estimation process. The above form of J_i requires that $\hat{\mathbf{x}}$ minimize J such that:

$$J = \sum_{i=1}^{m} J_i$$

The optimality conditions are:

$$g(\hat{\mathbf{x}}) = 0 = -\frac{1}{2} \sum_{i=1}^{m} \frac{\partial J_i}{\partial \mathbf{x}}\bigg|_{\mathbf{x}=\hat{\mathbf{x}}}$$

where

$$\left(-\frac{1}{2}\right)\frac{\partial J_i}{\partial \mathbf{x}} = \begin{cases} \dfrac{1}{\sigma_i}\dfrac{\partial h_i}{\partial \mathbf{x}} r_i(\mathbf{x}), & |r_i| \le a \\ \dfrac{1}{\sigma_i}\left(\dfrac{a}{|r_i|}\right)^{\frac{3}{2}}\dfrac{\partial h_i}{\partial \mathbf{x}} r_i(\mathbf{x}), & |r_i| > a \end{cases}$$

From this we can conclude that if $|r_i| > 0$, the effective standard deviation for the ith measurement is:

$$\sigma_i' = \sigma_i \frac{|r_i|^{\frac{3}{2}}}{a},$$

that is, bad measurements will effectively have large deviations of error. The information matrix associated with the above problem is given by:

$$\mathbf{F}(\mathbf{x}) = \sum_{i}^{m} \mathbf{F}_i(\mathbf{x}),$$

where

$$\mathbf{F}_i(\mathbf{x}) = \begin{cases} \dfrac{1}{\sigma_i^2}\dfrac{\partial h_i}{\partial \mathbf{x}}\left(\dfrac{\partial h_i}{\partial \mathbf{x}}\right)^T, & |r_i| \le a \\ \dfrac{1}{\sigma_i^2}\left(\dfrac{a}{|r_i|}\right)^{\frac{3}{2}}\dfrac{\partial h_i}{\partial \mathbf{x}}\left(\dfrac{\partial h_i}{\partial \mathbf{x}}\right)^T & |r_i| > a \end{cases}$$

The resulting iterative algorithm is given by:

$$\hat{\mathbf{x}}^{k+1} = \hat{\mathbf{x}}^k + \mathbf{F}^{-1}\left(\hat{\mathbf{x}}^k\right)\mathbf{g}\left(\hat{\mathbf{x}}^k\right).$$

In the context of pre- and postestimation analysis, robust estimation can be made quite effective. Basically all validated measurements will have a quadratic J_i function. Only suspected and raw measurements will have J_i functions of the type above. By extending this concept to parameter estimation, we can also identify large parameter errors. Figure 4.14 depicts a simplified flow chart for performing robust state and parameter estimation with pre-estimation bad data analysis.

4.10.6   SE for the Smart Grid Environment

These attributes are desirable in the development of future SE for smart grid computational tools:

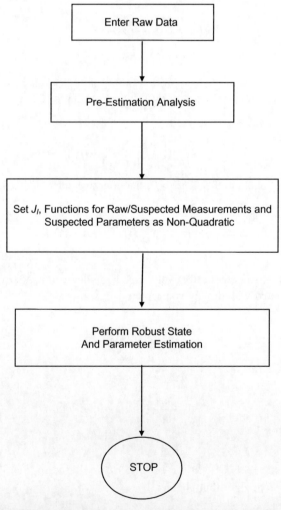

Figure 4.14. Flowchart for robust state estimation with pre-estimation analysis.

1. PMU-based SE includes the following steps suggested by Abur and Zhao [17]:

   a) Obtain PMU measurements at precisely controlled instants to estimate slow errors

   b) Measure for magnitude and angle of bus voltages (do not have to estimate angle)

   c) Measure the magnitude and angle of current

   d) Detect the state changes to reduce errors in building the bus admittance matrix

   e) Directly compute active and reactive power flows at the substation, or simply use voltage and current phasors to calculate active, reactive, and apparent power

In the new environment, the pattern of power flows in the network is less predictable than in the vertically integrated systems, in view of the possibilities associated with open access and the operation of the transmission network under energy market rules. Although reliability remains a central issue, the need for real-time network models us more important due to the new energy market-related functions which have to be added to new and existing EMS. These models are based on the results yielded by SE and are used in network applications such as optimal power flow, available transfer capability, voltage, and transient stability. The new role of SE and other advanced analytical functions in competitive energy markets is discussed by Shirmohammadi et al. [2]. Based on these network models, operators will be able to justify their technical and economic decisions, such as congestion management and the procurement for adequate ancillary services, and to identify potential operational problems related to voltage and transient stability [1].

4.10.7   Real-Time Network Modeling

Real-time models are built from a combination of snapshots of real-time measurements and static network data. The real-time measurements consist of analog measurements and the status of switching devices, whereas static network data corresponds to the system's parameters and basic substation configurations. Hence, the real-time model is a mathematical representation of the network's current conditions extracted at intervals from SE results. Ideally, SE should run at the scanning rate, for example, every two seconds, but due to computational limitations, most practical estimators run every few minutes or when a major change occurs.

4.10.8   Approach of the Smart Grid to State Estimation

SE is an important tool for detecting and diagnosing errors in measurement such as network error and/or device malfunctions. The technique is used in estimating voltage and power flow errors as a result of system parameters errors. The problem is formulated as one of the minimum $J(x)$ subject to different constraint conditions and solved by using a DC or AC power model for determining optimum performance. The formulation is reproduced as $J = (x - x^*)\, R(x - x^*)$ subject to varying constraints.

To solve this, a typical optimization technique could be employed with the following methodology and algorithm:

1. Convergence of the algorithm
2. Clearly defined problem location
3. Manage/modify noise server
4. Time and quality of measurements

Components suggested by SE researchers for the smart grid include:

1. Real PMU-based measurements to account for integrated RER/CG and protected load (stochastic) stored
2. A distributed SE around the different (partitioned) network configuration on appropriate assessment of data performed for subsequent assessment.

Figure 4.15 shows how PMU can be used to enhance the capability of the state estimator with the PMU schemes by adding phasor measurements with a linear step to traditional measurements and with a nonlinear iterative procedure. This approach is concerned with identifying major changes in the PMU in the SE, a final conclusion is that the data provide direct assessment of the state, which can significantly simplify the traditional SE and involves no change in existing SE. The changes could be in the techniques of solution. For example, the problem can be formulated as two subproblems, one composed only of SCADA measurements and another by PMU and SCADA. The first model is solved in the traditional way. The second can be solved as a local joint SE, which divides the system into locally observable islands. During the estimation process, based on the method of WLS, the measurements of PMU and SCADA

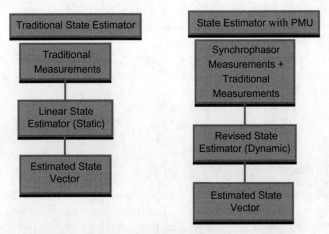

Figure 4.15. Scheme for state estimation with PMU.

can be mixed to correct the local state variables from the PMU. When the states of the PMU are reliable and accurate, the local estimator could be studied for omission. A system with enough PMU would give a direct measure of itself, or if a small number of PMUs exist, achieve further approximation in the estimate.

The hybrid of SE and PMU could generate a new data set for security assessment. Following this stochastic nature we can generate a predictive strategy for data analysis while suggesting real-time approximate control measures.

Conceptual visualization of an interconnected network can be used for vertically integrated utilities, pools, and ISOs. Ideally, the control area for which a specific control center is responsible is observable, although this is not always the case, since parts of it can be permanently or temporarily unobservable, for example, lower-voltage subnetworks. On the other hand, parts of the network outside the control area, which are normally unobservable, can be made observable by direct metering or data exchange.

Observable islands are handled with full SE including bad data [18, 19] analysis. SE can be extended to the rest of the system of interest through the addition of pseudomeasurements based on load prediction and generation scheduling. In executing SE for this augmented system, care must be taken to avoid corrupting the states estimated from telemetry data.

Hence, the state estimator is used to build the model for the observable part of the network and optionally to attach a model of the unobservable part. With adequate redundancy level, SE can eliminate the effect of bad data and allow the temporary loss of measurements without significantly affecting the quality of the estimated values. SE is mainly used to filter redundant data, to eliminate incorrect measurements, and to produce reliable state estimates, although to a certain extent it allows the determination of the power flows in parts of the network that are not directly metered. Not only will traditional applications, such as contingency analysis, optimal power flow, and dispatcher training simulation, rely on the quality of the real-time network model obtained via SE, but the new functions needed by the emerging energy markets will do so even more [2].

4.10.9 Dynamic State Estimation

The accuracy of the state estimator is very important because it feeds EMS functions, that is, voltage/angle stability, economic dispatch, security analysis, and so on. Today's power system is clearly more dynamic since both load and source vary with the introduction of RER as distributed resources. A technique to address this new dynamic system, dynamic state estimation (DSE), is currently being researched [17].

DSE incorporates advanced mathematical modeling, in which time stamping at instant k is noted, the state for $k + 1$ is predicted, and the errors are calculated to attain a single advanced time stamp. Information on one advance time stamp will help system operators make better decisions. The DSE module uses Holt's double exponential smoothing technique for predicting the state vector one time stamp ahead and the extended Kalman filter technique for filtering. To date, the usual SCADA data is utilized in lieu of PMU data and therefore the accuracy of the system is not very high.

4.10.10 Summary

Chapter 4 has discussed several performance tools for incorporation into smart grid design. The tools included voltage and angle stability and state estimation. Reviews of these methods were presented along with the research challenges.

REFERENCES

[1] M.H. Mickle and T.W. Sze. *Optimization in Systems Engineering*. Scranton, 1972.

[2] J.A. Momoh. *Electric Power System Application of Optimization*. Marcel Dekker, New York, 2001.

[3] G. Riley and J. Giarratano. *Expert Systems: Principles and Programming*. PWS Publisher, Boston, MA, 2003.

[6] A. Englebrecht. *Computational Intelligence: An Introduction*. John Wiley & Sons, 2007.

[7] M. Dorigo and T. Stuzle. *Ant Colony Optimization*. Massachusetts Institute of Technology, Cambridge, MA, 2004.

[6] P.K. Skula and K. Deb. "On Finding Multiple Pareto-optimal Solutions Using Classical and Evolutionary Generating Methods." *European Journal of Operational Research* 2007, 181, 1630–1652.

[7] C.W. Taylor. "The Future in On-Line Security Assessment and Wide-Area Stability Control." *IEEE Power Engineering Society* 2002, 1, 78–83.

[8] "Appendix B2: A Systems View of the Modern Grid-Sensing and Measurement." *National Energy Technology Laboratory* 2007.

[9] T. Bottorff. "PG&E Smart Meter: Smart Meter Program." NARUC Summer Meeting, 2007.

[10] D. Zhengchun, N. Zhenyong, and F. Wanliang. "Block QR Decomposition Based Power System State Estimation Algorithm." *ScienceDirect* 2005.

[11] A. Jain and N.R. Shivakumar. "Power System Tracking and Dynamic State Estimation."

[12] Y. Tamura, K. Sakamoto, and Y. Tayama. "Voltage Instability Proximity Index (VIPI) Based on Multiple Load Flow Solutions in Ill-Conditioned Power System," Sixth IEEE Conference on Decision and Control, vol. 3, pp. 2114–2119, August 2002.

SUGGESTED READINGS

A.G. Barto, W.B. Powell, D.C. Wunsch, and J. Si. *Handbook of Learning and Approximate Dynamic Programming*. IEEE Press Series on Computational Intelligence, 2004.

M. Dorigo and T. Stutzle. "The Ant Colony Optimization Metaheuristic: Algorithms, Applications and Advances." In F. Glover and G. Kochenberger, eds.: *Handboook of Metaheuristics*. Norwell, MA, Kluwer, 2002.

R.C. Eberhart and J. Kennedy. "A New Optimizer Using Particle Swarm Theory." In *Proceedings of the Sixth International Symposium on Micromachine and Human Science*, 1995; 39–43.

M. Gibescu, C-C Liu, H. Hashimoto, and H. Taoka. "Energy-Based Stability Margin Computation Incorporating Effects of ULTCs." In *IEEE Transactions on Power Systems* 2005, 20.

J.L. Marinho and B. Stott. "Linear Programming for Power System Network Security Applications." *IEEE Transactions on Power Apparatus and Systems* 1979, PAS-98, 837–848.

B. Milosevic and M. Begovic. "Voltage-Stability Protection and Control Using a Wide-Area Network of Phasor Measurements." *IEEE Transactions on Power Systems* 2003, 18, 121–127.

J. Momoh. *Electrical Power System Applications of Optimization.* CRC Press, 2008.

A.G. Phadhke. "Synchronized Phasor Measurments in PowerSystems." *IEEE Computer Applications in Power* 1993, 6, 10–15.

W.H. Zhange and T. Gao. "A Min-max Method with Adaptive Weightings for Uniformly Spaced Pareto Optimum Points." *Computers and Structures*, 2006, 84, 1760–1769.

L. Zhao and A. Abur. "Multiarea State Estimation Using Synchronized Phasor Measurements." *IEEE Transactions on Power Systems* 2005, 20, 2.

5

COMPUTATIONAL TOOLS FOR SMART GRID DESIGN

5.1 INTRODUCTION TO COMPUTATIONAL TOOLS

Previous work undertaken in the operation and research, systems engineering, and computer sciences communities to design various optimization and computational intelligence techniques has already been incorporated into large-scale grids; for example, the work in References 1–37 using artificial intelligence, heuristic and evolutional optimization to analyze optimal power flow, power flow, SE, stability, and unit commitment. This chapter underscores the importance of computational tools by discussing the answers to the following three questions:

1. Are the tools sufficient for modeling and accounting for adequate models of the system as it incorporates variability and randomness of RER?
2. Can the tools manage stochasticity and randomness in the system?
3. Can the tools address predictivity and the anticipatory nature of the problems encountered?

The classical optimization tools currently used cannot handle the adaptability and stochasticity of smart grid functions. Thus, the computational tools and techniques

Smart Grid: Fundamentals of Design and Analysis, First Edition. James Momoh.
© 2012 Institute of Electrical and Electronics Engineers. Published 2012 by John Wiley & Sons, Inc.

required are defined as a platform for assessment, coordination, control, operation, and planning of the smart grid under different uncertainties. We define the competitive schemes which are able to handle:

1. Inadequate models of the real world.
2. Complexity and large size of the problems which prohibit computation using computational intelligence.
3. Solution method employed by the operator which is incapable of being expressed in an algorithm or mathematical form, usually involves many rules of thumb, and is limited.
4. Decision-making by operator is based on fuzzy linguistics description.

We next discuss the concepts, algorithms, and some of the drawbacks of the proposed computational tools.

5.2 DECISION SUPPORT TOOLS (DS)

Decision support tools (DS) combining game theory, decision support systems, and analytical hierarchical processes (AHP) are used for computation of multiobjectives and risk assessment in smart grid planning and operations. Decision analysis (DA) is a powerful tool that makes a total uncertainty problem appear as a perfectly rational decision that is based on numerical values for comparing and yielding fast results. It looks at the paradigm in which an individual or group decision-maker contemplates a choice of action in an uncertain environment.

Importantly, the process relies on information about the alternatives. The quality of information varies from hard data to subjective interpretations, from certainty about decision outcomes (deterministic information) to uncertain outcomes represented by probabilities and fuzzy numbers. This diversity in type and quality of information about a decision problem calls for methods and techniques that can assist in information processing [7].

DA includes many procedures, methods, and tools for identifying, clearly representing, and formally assessing the important aspects of a decision situation. It computes the recommended course of action by applying the maximum expected utility action axiom to a well-formed representation of the decision, translating the formal representation of a decision and its corresponding recommendation into insights for the decision-maker(s). Multi-criteria decision analysis (MCDA), a form of DA, supports decision-maker(s) when a problem involves numerous conflicting evaluations. MCDA highlights these conflicts and derives the path to a compromise in a transparent process. Analytical hierarchical processing (AHP) is another form of MCDA.

DA support requires two stages: **Stage 1**: Evaluate the expected monetary value (EMV) from the profit and loss data and their associated probabilities. Draw the first decision flow tree which should yield a best decision based on the highest EMV and/or the lowest expected loss.

Stage 2: Consider the possibilities of sampling and accurate information and reevaluate the new EMV. Draw the new decision flow diagram and include the tree in stage 1. This should yield a best decision based on the highest EMV and/ or the lowest expected loss.

DA must be implemented with care. If the available data are inadequate to support the analysis, it is difficult to evaluate the effectiveness, which will lead to oversimplification of the problem. The outcomes of DA are not amenable to traditional statistical analysis. Strictly, by the tenets of DA, the preferred strategy or treatment is the one that yields the greatest utility (or maximizes the occurrence of favorable outcomes) no matter how narrow the margin of improvement. DA is particularly useful in handling multiobjective functions, or attaining several goals of the smart grid where risk and possibility are included.

5.2.1 Analytical Hierarchical Programming (AHP)

AHP is a decision-making approach that presents the alternatives and criteria, evaluates the trade-offs, and performs a synthesis to arrive at a final decision. AHP is especially appropriate for cases involving both qualitative and quantitative analyses. Figure 5.1 illustrates the AHP method.

AHP has found its widest application in multicriteria decision-making, in planning and resource allocation, and in conflict resolution [7]. In its general form, AHP is a nonlinear framework for carrying out both deductive and inductive thinking without use of the syllogism by considering several factors simultaneously, allowing for dependence and feedback, and making numerical tradeoffs to arrive at a synthesis

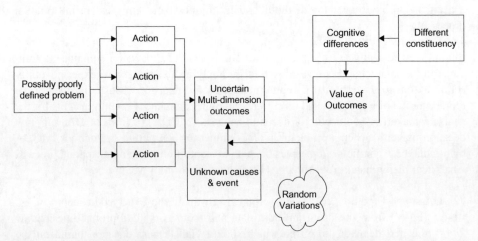

Figure 5.1. Analytical Hierarchical Programming (AHP).

or conclusion. The composite priorities of each alternative at the bottom level of a hierarchy may be represented as a multilinear form:

$$\sum_{i_1,\ldots,i_p} x_1^{i_1} x_2^{i_2} \ldots x_p^{i_p}$$

Consider a single term of this sum; for simplicity, denote it by $x_1, x_2 \ldots x_p$. The product integral is given by:

$$x_1 x_2 \cdots x_p = e^{\log x_1 x_2 \cdots x_p} = \prod_{i=1}^{n} e^{\log x_i} = e^{\sum_{i=1}^{n} \log x_i} \rightarrow e^{\int \log x(\alpha)\,d\alpha}$$

5.3 OPTIMIZATION TECHNIQUES

Optimization techniques consist of static and dynamic techniques for optimization, such as linear programming, mixed integer, dynamic programming, and so on, for development of smart grid optimization and planning activities (see Fig. 5.2). Linear programming, nonlinear mixed-integer programming (MIP), dynamic programming (DP) and Lagrangian relaxation methods are used for power system and operation issues, but they are limited for use in the smart grid due to the static network of the programs they can solve. They work better when computed in conjunction with DS tools and computational tool techniques. Below is a summary of the highlights of their formulation, implementation, and problem process.

5.4 CLASSICAL OPTIMIZATION METHOD

5.4.1 Linear Programming

Linear programming uses a mathematical model to describe the problem with linear objectives and linear constraints. The general structure of problems solved by this method is:

Maximize $c^T x$
s.t. $Ax \le b$
and $x_i \ge 0 \quad \forall_i \in \{1, n\}$

Assume that the model is statically linear. The process to achieve the global optimum uses Simplex-like techniques, variants of the interior point method, or integer programming. These methods are applicable to problems involving linear objective functions and linear constraints. The solution procedure includes:

1. Initialization Step: Introduce slack variables (if needed) and determine initial point as a corner point solution of the equality constraints.
2. At each iteration, move from the current basic feasible solution to a better, adjacent basic feasible solution.

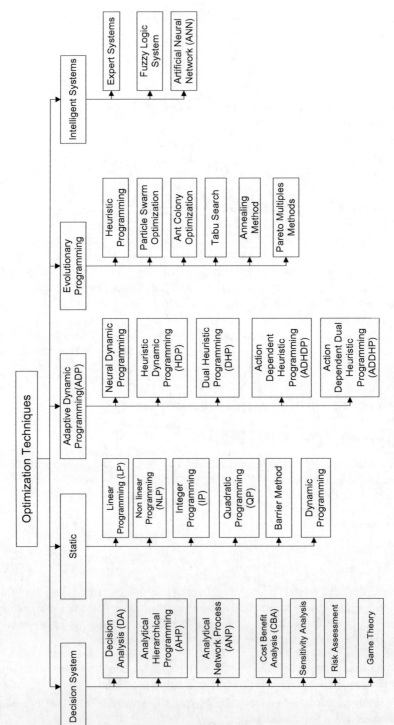

Figure 5.2. Optimization techniques.

3. Determine the entering basic variable: Select the nonbasic variable that, when increased, would increase the objective at the fastest rate. Determine the leaving basic variable: Select the basic variable that reaches zero first as the entering basic variable is increased.
4. Determine the new basic feasible solution.
5. Optimality Test and Termination Criteria: Check if the objective can be increased by increasing any non-basic variable by rewriting the objective function in terms of the nonbasic variables only and then checking the sign of the coefficient of each non-basic variable. If all coefficients are nonpositive, this solution is optimal; stop. Otherwise, go to the iterative.

Appling the linear programming method to the smart grid requires improving the it to accommodate the grid's stochasticity, predictivity, adaptivity, and randomness. The traditional linear method is confined to static problems and thus is insufficient for smart grid implementation.

5.4.2 Nonlinear Programming

Nonlinear programming (NLP) typically employs Lagrangian or Newtonian techniques for constrained and unconstrained optimization problems. The approach assumes that all objective functions are modeled as smooth and continuous functions. However, the mathematical response of the power system diverts from this assumption in many instances.

The general structure of problems solved by this method is:

Maximize f(y)
Subject to
$C_i \leq f_i(y) \leq D_i \quad \forall i \in \{1, n\}$

The procedure used in solving the NLP class of problems stems from identifying the feasibility of satisfaction. Convergence is due to sufficiency condition as given in work defined by extended KT conditions. Reference 36 provides a detailed formulation and methodology to solve the NLP class of problems. The sensitivity method, barrier method, and quadratic programming all feature in this class. Briefly stated, for all NLP:

1. Determine the initial feasible set based on investigation of extrema of the functions with or without constraints.
2. Check the optimality conditions.
3. Determine candidate solution for local or global optimum.
4. Perform further optimization and evaluate the optimal value to the objective function that satisfies the constraints.

This process may involve the application of Kuhn-Tucker (KT) and Karush-Kuhn-Tucker (KKT) first and second order necessary and sufficient conditions [36, 37]. It can be applied to functions as well as functional. NLP problems can be classified based on the nature of the objective functions and constraints.

The two challenges of NLP are its large computational burden and its limitation to static variables in the objectives and constraints. Moreover, NLP including QP often suffers if the approximations to the actual models for these devices are not sufficiently accurate.

Apart from being a very common form for many important problems, QP is also useful because many of the problems in a power system are often solved as a series of QP problems or Sequential Quadratic Programming (SQP).

Again, as in LP, the nonlinear programming classes are not suitable or designed to handle variability and predictivity. Therefore, research is needed to include this smart grid feature if they are to be used as computational tools. More work is needed to improve this technique to include the special requirement for the smart grid optimization process which requires adaptive, predictive, and stochastic algorithms.

5.4.3   Integer Programming

This is a special case of LP where all or some of the decision variables are restricted to discrete integer values, for example, where the discrete values are restricted to zero and one only, that is, yes or no decisions, or binary decision variables. The general structure of the MIP problem is:

$$\text{Maximize } P(x) = \sum_{j=1}^{n} c_j x_i$$

$$\text{Subject to the constraints: } \sum_{i=1}^{m} \left(\sum_{i=1}^{n} a_{ij} x_j \leq b_i \right)$$

And $x_j \geq 0 \ \forall \ j \in \{1, n\}$, and x_j is an integer $\forall_i \in \{1, I\}$.

The branch-and-bound procedure features:

1. Initialization: Set $P^* = -\infty$, where P^* is the optimal value of P.
2. Branching: This step involves developing subproblems by fixing the binary variables at 0 or 1, or choosing the first element in the natural ordering of the variables as the branching variable.
3. Bounding: For each of the subproblems, a bound can be obtained to determine the goodness of its best feasible solution. For each new subproblem, obtain its bound by applying the Simplex method to its LP relaxation and use the value of the P for the resulting optimal solution.
4. Fathoming: If a subproblem has a feasible solution, it should be stored as the first incumbent (the best feasible solution found so far) for the whole problem along with its value of P. This value is denoted P^*, which is the current incumbent for P.
5. Optimality Test: The iterative procedure stops when no subproblems remain. At this stage, the current incumbent for P is the optimal solution.

Pure integer or MIP problems pose a great computational challenge. While highly efficient LP techniques can enumerate the basic LP problem at each possible combination of the discrete variables (nodes), the problem lies in the astronomically large number of combinations to be enumerated. If there are N discrete variables, the total number of combinations becomes 2^N! The branch-and-bound technique for binary integer or reformulated MIP problems overcomes this challenge by dividing the overall problem into smaller and smaller sub-problems and enumerating them in a logical sequence. Note that the direct use of these techniques for solving the smart grid optimization problem will be limited, because they are generally static and are not designed for handling real-time and dynamic optimization problems.

5.4.4 Dynamic Programming

This approach was developed to solve sequential, or multistage, decision problems. Basically, it solves a multivariable problem by solving a series of single variable problems. This is achieved by tandem projection onto the space of each of the variables. In other words, it projects first onto a subset of the variables, then onto a subset of these, and so on. It is a candidate optimization technique for handling time variability and noise in the objective and constrained optimization problem.

Two common techniques derived from Bellman principles are

1 Backward and forward recursion or table look up
2 Calculus based on composition

Both techniques have the drawback of the curse of dimensionality and thus are not suitable for large, complex power systems.

5.4.5 Stochastic Programming and Chance Constrained Programming (CCP)

Stochastic programming solves LP problems where the uncertainty assumption is so badly violated that the same parameters must be treated explicitly as random variables. The two ways to handle LP with variability are:

1. Stochastic programming (SP)
2. Chance-constrained programming (CCP)

SP requires all constraints to hold with probability whereas CCP permits a small probability of validating any functional constraint.

Formulation of Stochastic Programming. Consider max $Z = \sum c_j x_j$ between the limits 1 and n. Because we can replace Z by some deterministic function, we pose it as $E(Z) = \sum cjxj$ between the limits 1 and n. Similarly, the functional constraints $\sum a_{ij} x_j \leq b_j$ for $i = 1, 2, 3, 4 \ldots m$. The solution requires feasibility requirements of the constraints for all possible combinations of parameters in a_{ij}, b_j which must be assigned to x_j.

The problem is solved in two categories:

1. One-stage problem: Here, a_{ij}, b_j are random variables and mutually independent where both a_{ij}, b_j must satisfy: $\sum \max a_{ij} x_j \leq \min b_j$
2. Multistage problem: Here, x_j have 2 or more points in the time stage. x_j are a firststage variable, others are secondstage variables, and so on. This method is based on first obtaining the first time and the subsequent time depending on what happens in the preceding stages. Therefore, SP for multistage problems is more adjustable for later stages decisions based on unfolding encounters.

When some of the parameters of the model are random variables (RVs), the SP formulation requires that all the functional constraints must hold for all possible combination of values for RV parameters. However, the CCP formulation requires only that each constraint be held for most of the combinational period. This formulation replaces the original LP constituents of $\sum a_{ij} x_j \leq b_j$ by $P\{\sum a_{ij} x_j \leq b_j\} \geq \alpha_1$ where α_1 are specified between zero and one. They are normally chosen to be reasonably close to one. Therefore, a non-negative solution $(x_1, x_2, x_3, \ldots, x_n)$ is considered feasible if and only if $P\{\sum a_{ij} x_j \leq b_j\} \geq \alpha_1$ for $n = 1, 2, 3 \ldots n$. Each contingency probability $1 - \alpha_1$ represents the allowable risk such that the RV will take on value stated as $a_{ij} x_j \leq b_j$ for $j = 1, 2, 3, \ldots n$. Thus, the best objective is to select the best non-negative solution that will likely satisfy each of the original constraints when RV (a_{ij}, b_i, c_i) take on the values.

In general, the procedure for computing CCP, LP includes all a_{ij} parameters which must be correlative so that b_i, c_i are RV probability distribution of b_i and c_j is slightly independent of b_j (where $j = 1, 2, 3, 4 \ldots n$; for $i = 1, 2, 3, \ldots m$).

SP is positioned to solve network, transportation, and power grid problems with uncertainties, randomness, and noise. To date, work on SP is able to handle small-sized systems, but is still unable to account for randomness and predictivity.

5.5 HEURISTIC OPTIMIZATION

Future work that accounts for the predictive and stochastic nature of the smart grid involves:

1. Modeling components to account for predictivity and stochasticity
2. Selecting new optimization methods such as adaptive dynamic programming (ADP)

1. The security of power systems is affected by various contingencies which may move the system from normal and alert to emergency states. Their impacts are studied with the optimization tools under development. The limitations to be resolved include the fact that a model cannot account for uncertainty in load, central location, and sources.
2. Selectivity of contingency sets appropriate for study (in real time for offline study).
3. Accounting for stability of the solution and model price and economic marginal benefits of new RER.

A new optimization strategy is proposed:

1. Model a new objective function to account for customer and welfare power uncertainty.
2. Model RER to account for stochasticity and variability.
3. Update the model and simulate.
4. Define new system components' performance in time.
5. Following this, solve the probabilistic load flow for base case studies and include new contingency set to define violations.
6. Select appropriate optimization technique which may include heuristic program technique and hybrid or ADP, methods that define the impact of predictivity and stochasticity as optimum.

5.5.1 Artificial Neural Networks (ANN)

ANN are based on the natural genetics of the brain. Common techniques include back and forward propagation techniques. ANN have the ability to classify and recognize patterns in a large quantity of data through training and tuning of the algorithm. The key element of this paradigm is the novel structure of the information processing system, which comprises a large number of highly interconnected processing elements (neurons) working in unison to solve specific problems. ANN learn by example. They are configured for a specific application, such as pattern recognition or data classification, through a learning process.

ANN, with their remarkable ability to derive meaning from complicated or imprecise data, can be used to extract patterns and detect trends that are too complex to be noticed by either humans or other computer techniques. A trained neural network can be thought of as the expert in the category of information it has been given to analyze. The expert provides projections, given new situations of interest and answers the "what if" questions.

Other advantages of ANN include:

1. Adaptive learning: Ability to learn how to do tasks based on the data given for training or initial experience.
2. Self-organization: Creates its own organization or representation of the information received during learning period.
3. Real-Time Operation: Computations may be carried out in parallel; special hardware devices are being designed and manufactured to exploit this capability.
4. Fault Tolerance via Redundant Information Coding: Partial destruction of a network leads to the corresponding degradation of performance.

Back-propagation nets are probably the most well-known and widely applied of the neural networks in use today. The net is a perceptron with multiple layers, a different threshold function in the artificial neuron, and a more robust and capable learning rule. A unit in the output layer determines its activity in two steps:

Step 1. It computes the total weighted input x_j, using the formula: $X_j = \sum y_i W_{ij}$

where y_i is the activity level of the jth unit in the previous layer and W_{ij} is the weight of the connection between the ith and the jth unit.

Step 2: The unit calculates the activity y_j using some function of the total weighted input. Typically we use the sigmoid function:

$$y_j = \frac{1}{1 + e^{-x_j}}$$

Once the activities of all output units have been determined, the network computes the error E, which is defined by the expression:

$$E = \frac{1}{2} \sum_i (y_i - d_i)^2$$

where y_j is the activity level of the jth unit in the top layer and d_j is the desired output of the jth unit.

The back-propagation algorithm consists of four steps:

Step 1: Compute how fast the error changes as the activity of an output unit is changed. This error derivative (*EA*) is the difference between the actual and the desired activity.

$$EA_j = \frac{\partial E}{\partial y_j} = y_j - d_j$$

Step 2: Compute how fast the error changes as the total input received by an output unit is changed. This quantity (*EI*) is the answer from Step 1 multiplied by the rate at which the output of a unit changes as its total input is changed.

$$EI_j = \frac{\partial E}{\partial x_j} = \frac{\partial E}{\partial y_j} \times \frac{dy_j}{dx_j} = EA_j y_j (1 - y_j)$$

Step 3: Compute how fast the error changes as a weight on the connection into an output unit is changed. This quantity (*EW*) is the answer from Step 2 multiplied by the activity level of the unit from which the connection emanates.

$$EW_{ij} = \frac{\partial E}{\partial W_{ij}} = \frac{\partial E}{\partial x_j} \times \frac{\partial x_j}{\partial W_{ij}} = EI_j y_i$$

Step 4: Compute how fast the error changes as the activity of a unit in the previous layer is changed. To compute the overall effect on the error, add all of these separate effects on output units.

$$EA_i = \frac{\partial E}{\partial y_i} = \sum_j \frac{\partial E}{\partial x_j} \times \frac{\partial x_j}{\partial y_i} = \sum_j EI_j W_{ij}$$

Power system applications include: adaptive control, fault detection and classification, network reconfiguration, voltage stability assessment, and transient stability assessment. These functions require real-time data for monitoring and control. The deployment of ANN as a computational platform will facilitate the assessment of smart grid performance for each of the above applications.

5.5.2 Expert Systems (ES)

ES are used as a method of optimization that relies on heuristic or rule-driven decision-making. They are sometimes used for fault diagnosis with prescription for corrective actions. While the expert system/computer application performs a task that would otherwise be performed by a human, the method is only as reliable as the designed engineering rule-base. Figure 5.3 depicts the components.

Expert systems have several advantages over human experts, including: increased availability and reliability, lower cost and response time, increased confidence in decision-making ability via provision of clear reason for a given answer. Its uses for the power system include optimal load shedding, resource allocation such as VAr, discrete control (series capacitors, ULTCs, and so on), and economic dispatch. The operator-assisted functions in the management of the smart grid will benefit from the use of this knowledge-based system or its hybrid with ANN, or other variations of computational/heuristic optimization techniques. Some of the problems can be reformulated using expert systems, real time data, and hybrid ANN.

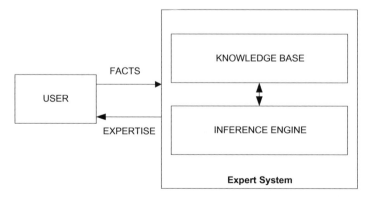

Figure 5.3. Fundamental components of an expert system.

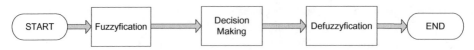

Figure 5.4. Simplified block diagram of fuzzy logic approach.

Fuzzy logic is a superset of conventional logic that has been extended to handle the concept of partial truth, which are the truth values between "completely true" and "completely false." Special membership functions transform uncertainty in data to a crisp form for analysis. Fuzzy-logic control is often used to determine whether the process variables are within acceptable tolerances. A fuzzy set is a generalization of ordinary sets that allows assigning a degree of membership for each element to range from (0, 1) interval. Figure 5.4 depicts the simplified block diagram of the fuzzy logic approach.

A fuzzy system combines and applies sets with fuzzy rules to problems with overall complex nonlinear behavior. In smart grid design, many real-time decisions require the attribute of fuzziness. The design of automatic generation control (AGC), SSA, and SE will therefore be used for proposed smart grid functions and control.

5.6 EVOLUTIONARY COMPUTATIONAL TECHNIQUES

Based on natural genetics, EC solve combinatorial optimization problems. The techniques in this category, including particle-swarm, ant-colony, genetic algorithms (GA), and artificial intelligence, learn or adapt to new situations, generalize, abstract, discover, and associate. The evolutionary algorithms use a population of individuals. An individual is referred to as a chromosome which defines the characteristics of individuals in the population. The characteristic of each individual is termed a gene. Individuals with the best survival capabilities have the best chance to reproduce [7]. Offspring are generated by combining parts of the parents, a process referred to as crossover. Each individual in the population can also undergo mutation which alters some of the alleles of the chromosome.

5.6.1 Genetic Algorithm (GA)

GA mimics biological evolution such that the elements in the algorithm are synonymous with genetic system terminology. Figure 5.5 from References 7 and 8 shows a typical GA in which offspring are produced from selected parents, modified through crossover or mutation, and evaluated to find the fittest offspring. They are placed in the population to become parents while the unfit offspring are discarded. The process can be repeated until a suitable offspring or solution is created.

GA has been applied to power system expansion and structural planning, operation planning, and generation, transmission, and distribution operation and analysis for VAr planning requiring real-time operation with uncertainty and randomness.

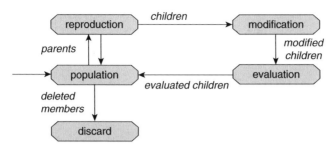

Figure 5.5. Typical GA cycle.

5.6.2   Particle Swarm Optimization (PSO)

(PSO) is a population-based stochastic optimization technique developed by Eberhart and Kennedy in 1995. It uses a simple mechanism that mimics the social behavior of bird flocking and fish schooling to guide the particles' search for globally optimal solutions.

Individuals in a particle swarm follow a very simple behavior: to emulate the success of neighboring individuals and their own successes. The collective behavior that emerges is that of discovering optimal regions of a high dimensional search space [5]. The population of the study is referred to as the population and each individual is termed a particle. The equation shows that for each time, t, the position of the particle x_i is altered by adding a velocity, $v_i(t)$:

$$x_i(t+1) = x_i(t) + v_i(t+1)$$

The velocity adjusts the particle based on both the particle and its neighbors within the swarm, which is calculated as:

$$v_{ij}(t+1) = v_{ij}(t) + c_1 r_{1j}(t)[y_{ij}(t) - x_{ij}(t)] + c_2 r_{2j}(t)[\hat{y}_j(t) - x_{ij}(t)]$$

where $v_{ij}(t)$ is the velocity of particle i in dimension $j = 1, \ldots, n$, $x_{ij}(t)$ is the position of particle i in dimension j at time step t, c_1 and c_2 are the positive acceleration constants used to scale the contribution of the cognitive and social components, respectively (discussed in Section 16.4), and $r_{1j}(t)$, $r_{2j}(t)$ are the random values in the range (0, 1) sampled from a uniform distribution which introduce a stochastic element to the algorithm.

Two fundamental variants of PSO are Local Best (lbest) and Global Best (gbest). The lbest provides a local best, utilizing a ring network topology and a subset of the swarm as the neighborhood of the particle. The gbest presents the global best where the neighborhood for the particle taken to be the swarm.

5.6.3   Ant Colony Optimization

ACO is a class which is applied to combinatorial optimization problems. The essential trait of ACO algorithms is the combination of a priori information about the structure of a promising solution with a posteriori information about the structure of previously obtained good solutions [6, 7]. ACO uses computational concurrent and asynchronous agents termed a colony of ants that move through states of the problem corresponding to the partial solutions. The measurement generally involves a stochastic local decision policy based on the two parameters, trails and attractiveness. The pheromone information will direct the search of the future. During the construction phase or upon completing the solution, the ant evaluates the solution and modifies the trail value of the components used in its solution.

An ACO algorithm includes trail evaporation and daemon actions. The trail evaporation action decreases all trail values over time, in order to avoid unlimited accumulation of trails over some component. Conversely, daemon actions implement centralized actions which cannot be performed by single ants, such as the invocation of a local optimization procedure, or the update of global information used to decide whether to bias the search process from a nonlocal perspective [7]. The process is:

> **begin**
> Intialize $\tau_{ij}(0)$
> Let $t = 0$;
> Set location, n_k = origin node;
> **while** stopping criteria is not true
> **for** $k = 1, \ldots, n_k$
> // Construct a path, $x^k(t)$;
> $x^k(t) = 0$;
>> **for** destination node is reached;
>> remove loops from $x^k(t)$;
>> calculate $f(x^k(t))$;
> **end**
> **end**
> **for** each link(i, j) of the graph
> //pheromone evaporation;
> $\tau_{ij}(t) = (1 - \rho)\tau_{ij}(t)$
> **end**
> **for** $k = 1, \ldots, n_k$
>> **for** link(i, j) of $x^k(t)$
>> $$\Delta\tau^k = \frac{1}{f\left(x^k\left(t \lim_{\delta x \to 0}\right)\right)}$$
>> $$\tau_{ij}(t+1) = \tau_{ij}(t) + \sum_{k=1}^{n_k} \Delta\tau_{ij}^k(t)$$
> **end**

end

$t = t + 1;$

end

return $x^k(t)$ for smallest $f(x^k(t))$.

5.7 ADAPTIVE DYNAMIC PROGRAMMING TECHNIQUES

ADP incorporates time dependency of the deterministic or stochastic data required for the future. ADP is also termed reinforcement learning, adaptive critics, neural-dynamic programming, and approximate dynamic programming [6, 7] (see Fig. 5.6). ADP consider the optimization over time by using learning approximation to handle problems that severely challenge conventional methods due to their very large scale and lack of sufficient prior knowledge. ADP overcomes the curse of dimensionality in DP. ADP is designed to maximize the expected value of the sum of future utility over all future time periods:

$$Maximize \sum_{k=0}^{\infty} (1+r)^{-k} U(t+k)$$

Some common nomenclature in ADP is:

$\underline{u}(t)$: Action vectors

$U(t)$: Utility which the system is to maximize

$\underline{X}(t)$: Sensor inputs

r: Discount rate or interest rate (needed only in infinite-time-horizon problems)

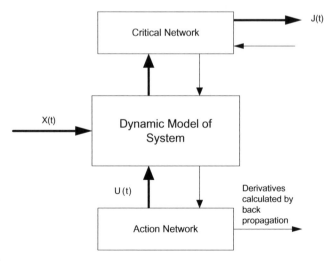

Figure 5.6. Structure of adaptive dynamic programming (ADP) system.

<>: Denote the expectation value

J: Secondary or strategic utility function

$R(t)$: Complete state description of the plant to be controlled at time t

A: Action network

F_W_{ij}: Derivatives of error with respect to all weights W_{ij}

Adaptive critic designs (ACDs) use neural networks based designs for optimization over time using combine concepts of reinforcement learning and approximate dynamic programming.

ACDs use two neural networks, the critic and action networks, to solve the Hamilton-Jacobi-Bellman equation of optimal control. The critic network approximates the cost to-go function J of Bellman's equation of dynamic programming and is referred to as the heuristic dynamic programming (HDP) approach in ACDs.

The action network provides optimal control to minimize or maximize the cost-to-go function J.

ACD use combined concepts of reinforcement learning and approximate dynamic programming. Two neural networks, the critic and action networks, are used to solve the Hamilton-Jacobi-Bellman equation of optimal control. The critic network approximates the cost-to-go function J of Bellman's equation of dynamic programming, termed the HDP approach, while the action network provides optimal control to minimize or maximize the cost-to-go function J.

Several critic designs have been proposed based on dynamic programming:

- Heuristic dynamic programming (HDP) adapts a critic network whose output is an approximation of $J(R(t))H$:

$$J(R(t)) = \underset{u(t)}{Max}(U(R(t), u(t)) + \frac{<J(R(t+1))>}{1+r}$$

- Dual Heuristic programming (DHP) adapts a critic network whose outputs represent the derivatives of $J(R(t))$:

$$J(R(t)) = (U(R(t), u(t)) + <J(R(t+1))> - U_0$$

$$\lambda_i(R(t)) = \frac{\partial J(R(t))}{\partial R_i(t)} = \frac{\partial}{\partial R_i(t)} U(R(t), u(R(t))) + <\frac{\partial J(R(t+1))}{\partial R_i(t)}>$$

$$= \frac{\partial J(R(t), u(t))}{\partial R_i(t)} + \sum_j \frac{\partial U(R, u)}{\partial u_j} \cdot \frac{\partial u_j R(t)}{\partial R_i(t)} + \sum_j \left\langle \frac{\partial J(R(t+1))}{\partial R_i(t+1)} \cdot \frac{\partial R_j(t+1)}{\partial R_i(t)} \right\rangle$$

$$+ \sum_{j,k} \left\langle \frac{\partial J(R(t+1)}{\partial R_j(t+1)} \cdot \frac{\partial R_j(t+1)}{\partial u_k(t)} \cdot \frac{\partial u_k(t)}{\partial R_j(t)} \right\rangle$$

- Globalized DHP (GDHP) adapts a critic network whose output is an approximation of $J(R(t))$ to minimize errors in the implied derivatives of J. GDHP tries to combine the best of HDP and DHP.

HDP intends to break down through very slow learning as the problem size grows, however, DHP is more difficult to implement. The three methods all yield action-independent critics, and there are also ways to adapt a critic network that inputs $R(t)$ and $u(t)$ (ADHDP, ADDHP).

Action dependent heuristic dynamic programming (ADHDP or Q-learning) adapts a critic network, $\underline{J}'(R(t),u(t),W)$, which attempts to approximate J' as defined in equations:

$$J'(R(t), u(t)) = U(R(t), u(t)) + \frac{<J(R(t+1))>}{1+r}$$

$$J'(R(t), u(t)) = U(R(t), u(t)) + \underset{u(t+1)}{Max} \frac{<J'(R(t+1), u(t+1))>}{1+r}$$

5.8 PARETO METHODS

Pareto analysis is used for solving the multiobjective optimization problem:

$$\min_{x \in N} F(x)$$

where $F(x) = (f_1, f_2, \ldots, f_m)^T$

s.t

$g(x) = 0$

$h(x) \leq 0$

$a \leq x \leq b$

The goal is to seek Pareto-optimal solutions through various approaches. Due to the conflicting nature of the objective functions based on the concepts of convergence and diversity, solutions to the above problem are both multiple and Pareto [3] optimal. By definition, a feasible solution X^* is Pareto optimal if there is no other improved feasible point X such that $f_k(X) f_k(X^*)$ with strict inequality for at least one condition.

Two methods for solving such problems are:

- One-at-a-time strategy: A multiobjective optimizer may be applied repeatedly with the goal each time of finding one single Pareto-optimal solution. These are classical generating multiobjective optimization methods which can be static or deterministic and which use such an iterative scalarization scheme of standard procedures, such as weighted-sum, ε-constraint method, trade-off method, or the min-max method [5, 6]. The drawbacks include inefficiency as well as the difficulty in maintaining diversity in the objective space. The solution to every subproblem involves contending with infeasible region and local optimums (which may feature in every step of the solution). Overcoming such difficulties requires an optimization algorithm to learn how to solve the problem independently each time for a different initial point [2].
- Simultaneous strategy: This approach utilizes EA due to the population or archive-based approach to facilitate a parallel search. The efficiency of this method is greatly improved due to the reduced need for multiple applications of

TABLE 5.1. Hybridized Computational Intelligence Tools for Smart Grid Analysis and Design.

	Fuzzy	GA	AOC	ANN	EP	PSO	ADP
Fuzzy		X	X	X	X	X	X
Genetic Algorithm (GA)	X			X	X		X
Ant colony optimization (ACO)	X						X
Artificial Neural Networks (ANN)	X	X			X		X
Evolution programming (EP)	X	X		X			X
Particle Swamp Optimization (PSO)	X	X		X	X		
Adaptive Dynamic Programming (ADP)	X		X			X	

the optimization. Critical information (including the information needed to solve a subproblem) is shared among the population through exchange (or recombination).

Pareto can be used to augment static optimization techniques.

5.9 HYBRIDIZING OPTIMIZATION TECHNIQUES AND APPLICATIONS TO THE SMART GRID

To facilitate the computational methods we incorporate hybrid of intelligent systems such as immunized-neuro systems, immunized-swarm systems, neuro-fuzzy systems, neuro-swarm systems, fuzzy-PSO systems, fuzzy-GA systems, and neuro-genetic systems. These hybrids (Table 5.1) are necessary to harness the advantages of the various paradigms of computational intelligence [7] for addressing the smart grid functions and capability to achieve self-learning capability.

These advanced techniques have a host of characteristics which make them viable for application to the smart grid environment, such as handling the nonlinear modeling of time varying dynamics of the power system, and handling dynamics and stochasticity. These techniques will be utilized for FACTS controllers, learning for systems, training sets for voltage profile, unit commitment, discrete adjustment of controls, power system planning, and parallel computing.

5.10 COMPUTATIONAL CHALLENGES

The computational challenges associated with utilizing advanced tools involve their optimal selection based upon the specific application and location. Data availability is another challenge, enforcing the need to develop sensor and communication technology to facilitate the acquisition of real-time or just-in-time data.

Some of the key questions to be answered are:

- **Software/hardware:** Is the tool to be developed software only or is some hardware required for integration and implementation?

- **Integration:** How will software/hardware be integrated into the system? Will it replace an existing tool or be integrated into an existing package/tool?
- **Location of installation**: Where will a tool be located and used; for example, LAN or WAN? Will it be necessary at all generating points or load points, at substation levels, or at the customer level?
- **Robustness:** How will the tool be implemented for ease of application and robustness?
- **Sensitivity:** What degree of sensitivity is required for satisfactory functioning?
- **Standards:** Have existing standards been identified or developed for utilization of the tool in similar environments?

In practice, the tools will be used by engineers and operators who lack knowledge and training. Thus, the advanced tools must be readily interpretable, user friendly, and self-teaching.

An expanded power system, deregulation (liberalization), and environmental concerns produce new challenges for operation, control, and automation. Power system models used for intelligent operation and control are highly dependent on the task purpose. Smart solutions to economic, technical (secure, stable, and good power quality), and environmental goals, must incorporate forecasting of load, price, and ancillary services; penetration of new and renewable energy sources; bidding strategies of participants; system planning and control; operating decision-making under missing information; increased distributed generations and demand response; tuning controller parameters in different operating conditions, and so on.

Risk management and financial management in the electric sector attempt to find the ideal trade-off between maximizing the expected returns and minimizing the risks associated with investment. Computational intelligence (CI) is a new tool for solving the complex problems posed by smart grid technology. For example, heuristic optimization techniques have been combined with elements of nature-based methods or those with a foundation in stochastic and simulation approaches. Research occurring in the field of intelligent system technologies is using digital signal processing techniques, expert systems, artificial intelligence, and machine learning. Developing solutions with these and other such tools offers two major advantages: development time is much shorter, and the systems are relatively insensitive to nois and/or missing data/information known as uncertainty. Due to environmental, rights-of-way, and cost problems, there is increased interest in better utilization of the capacity available in both bundled and unbundled power systems.

5.11 SUMMARY

Chapter 5 has summarized the classical optimization techniques and computational methods currently being applied in legacy and smart grid design, planning, and operations. The Pareto methods used in grid design were described. A discussion of computational challenges highlighted several problems faced by researchers and engineers in determining the best mitigation and solution approaches.

REFERENCES

[1] H.W. Dommel and W.F. Tinney. "Optimal Power Flow Solutions," *IEEE Transactions On Power Apparatus and Systems*, 1968, PAS.87, 1866–1876.

[2] W.L. Winston. *Operations Research: Applications and Algorithms*, Boston: Duxbury, 1987.

[3] J.A. Momoh. *Electric Power System Application of Optimization*, New York: Marcel Dekker, 2001.

[4] Y.L. Abdel-Magid, M.A. Abido, and A.H. Mantaway. "Robust Tuning of Power Systems Stabilizers in Multimachine Power Systems," *IEEE Transactions on Power Systems* 2000, 15, 735–740.

[5] M. A. Abido. "Robust Design of Multimachine Power System Stabilizers Using Simulated Annealing." *IEEE Transactions on Energy Conversions* 2000, 15, 297–304.

[6] R.A.F. Saleh and H.R Bolton. "Genetic Algorithm-aided Design of a Fuzzy Logic Stabilizer for a Superconducting Generator," *IEEE Transactions on Power Systems* 2000, 15 1329–1335.

[7] P. Zhang and A.H. Coonick. "Coordinated Synthesis of PSS Parameters in Multi-machine Power Systems Using The Method of Inequalities Applied to Genetic Algorithms," *IEEE Transactions on Power Systems* 2000, 15, 811–816.

[8] Y.L. Abdel-Magid, M.A. Abido, S. Al-Baiyat, and A.H. Mantawy. "Simultaneous Stabilization of Multimachine Power Systems Via Genetic Algorithms," *IEEE Transactions on Power Systems* 1999, 14, 1428–1439.

[9] A.L.B. do Bomfim, G.N Taranto, and D.M Falcao. "Simultaneous Turning of Power System Damping Controllers Using Genetic Algorithms," *IEEE Transactions on Power Systems* 2000, 15, 163–169.

[10] J. Wen, S. Cheng, and O.P. Malik. "A Synchronous Generator Fuzzy Excitation Controller Optimally Designed With a Genetic Algorithm," *IEEE Transactions on Power Systems* 1998, 13, 884–889.

[11] M.A. Abido and Y.L. Abdel-Magid. "Hybridizing Rule-based Power System Stabilizers With Genetic Algorithms," *IEEE Transactions on Power Systems* 1999, 14, 600–607.

[12] R. Asgharian and S.A. Tavakoli. "A Systematic Approach to Performance Weights Selection in Design of Robust H/sup/spl infin/PSS Using Genetic Algorithms," *IEEE Transactions on Energy Conversion* 1996, 11, 111–117.

[13] K.Y. Lee and J.G. Vlachogiannis. "Ant Colony Optimization For Active/reactive Operational Planning," World Congress, vol. 16, 2005.

[14] M.R. Kalli, I. Musirin, and M.M. Othman. "Ant Colony Based Optimization Technique for Voltage Stability Control," Proceedings of the 6th WSEAS International Conference on Power System, 2006.

[15] M.R. Kalli, I. Musirin, and M.M. Othman. "Optimal Transformer Tap Changer Setting For Voltage Stability Improvement," *International Journal of Power, Energy and Artificial intelligence* 2009, 1, 89–95.

[16] M.F. Mustafar, I. Musirin, M.R. Kalli, and M.K. Jebatan. "Ant Colony Optimization (ACO) Based Technique For Voltage Control and Loss Minimization Using Transformer Tap Setting," Proceedings of Research and Development 2007, 1–6, December 2007.

[17] H. Yoshida, K. Kawata, Y. Fukuyama, S. Takayama, and Y. Nakanishi. "A Particle Swarm Optimization For Reactive Power and Voltage Control Considering Voltage Security Assessment," *IEEE Transactions on Power Systems* 2000, 15.

[18] K.S. Pandya and S.K. Joshi. "A Survey of Optimal Power Flow Methods," *Journal of Theoretical and Applied Information Technology*, 450–458.

[19] P.A. Jeyanthy and D. Devaraj. "Optimal Reactive Power Dispatch For Voltage Stability Enhancement Using Real Coded Genetic Algorithm," *International Journal of Computer and Electrical Engineering* 2010, 2, 734–740.

[20] A.P. Alves da Silva and P.J. Abrao. "Applications of Evolutionary Computation in Electric Power Systems," IEEE, 2002.

[21] J.G. Vlachogiannis and J. Ostergaard. "Reactive Power and Voltage Control Based On General Quantum Genetic Algorithm," *Expert Systems with Applications* 2009, 36, 6118–6126.

[22] H. Ying-Yi and L. Yi-Feng. "Reactive Power Control in Distribution System Using Genetic Algorithms," Proceeding of the 14th International Conference on Intelligent System Application to Power System, 2007, 44–50.

[23] R. Shivakumar, R. Lakshmipathi, and Y. Suresh. "Implementation of Bio Inspired Genetic Optimizer in Enhancing Power System Stability," *IACSIT International Journal of Engineering and Technology* 2010, 2, pp. 263–268.

[24] B. Mahdad, T. Bouktir, and K. Srairi. "Genetic Algorithm and Fuzzy Rules Applied to Enhance the Optimal Power Flow with Considerations of FACTS," *International Journal of Computational Intelligence Research* 2008, 4, 229–238.

[25] A. Salami. "Control of Capacitors in Distribution Networks Using Neural Network Based On Radial Basis Function," Proceeding of the 5th IASME/WSEAS International Conference on Energy and Environment, 2010.

[26] M.A.H. El-Sayed. "Artificial Neural Network Reactive Power Optimization," *Neurocomputing*, 1998.

[27] L. Chen-Ching and T. Kevin. "An Expert System Assisting Decision-making of Reactive Power/Voltage Control," *IEEE Transactions on Power Systems* 2007, 1, 195–201.

[28] A.K. Tajfar, M. Iravani, and R.K. Zare. "Cordinated Fuzzy Logic Voltage/Var Controller in Distribution Networks," IEEE International Conference on Industrial Technology, 2008, 1–5.

[29] R. Shivakumar, V. Miranda, and P. Calisto. "A Fuzzy Inference System to Voltage/Var Control in Distribution Management System," Proceedings of 14th PSCC, Seville, 2002, vol. 1, 1–6.

[30] A.M.A. Haidar, A. Mohamed, and A. Hussain. "Vulnerability Control of Large Scale Interconnected Power System Using Neuro-fuzzy Load Shedding Approach," *Expert Systems with Applications* 2010, 10, issue 4.

[31] S.A. Karzalis, A.G. Barkitzis, and V. Petridis. "A Genetic Algorithm Solution of the Unit Commitment Problem," *IEEE Transactions* 1996, PWRS 11, 83–90.

[32] G.B. Sheble and T.T. Maifeld. "Unit Commitment by Genetic Algorithms and Expert System," EPSR 1994, 30, 115–121.

[33] D. Dasgupta and D.R. McGregor. "Thermal Unit Commitment Using Genetic Algorithms," *IEE Proceedings* 1994, Part C 141 5, 459–465.

[34] R. Meziane, Y. Massim, A. Zeblah, A. Ghoraf, and R. Rahli. "Reliability Optimization Using Ant Colony Algorithm Under Performance and Cost Constraints," Electric Power Systems Research, 2005.

[35] M. Sasson. "Non-linear Programming Solutions For Load Flow, Minimum Loss, and Economic Dispatching Problems," *IEEE Transactions on Power Apparatus and Systems* 1969, Pas-88.

[36] K.Y. Lee, Y.M. Park, and J.L. Ortiz. "A United Approach to Optimal Real and Reactive Power Dispatch," *IEEE Transactions on Power Systems* 1985, PAS-104, 1147–1153.

[37] H.W. Dommel and W.F. Tinney. "Optimal Power Flow Solutions," *IEEE Transactions on Power Apparatus and Systems*, 1968, PAS.87, 1866–1876.

6

PATHWAY FOR DESIGNING SMART GRID

6.1 INTRODUCTION TO SMART GRID PATHWAY DESIGN

The design of the smart grid involves the coupling of tools, technologies, and techniques [14]. An introduction to the tools and technologies which are required for the computational tool to capture the attribute of smart grid is given in this chapter. We will develop an algorithm which will support the computational tools for smart grid development. Due to the increased availability of real time or near-real time data and signals, it is necessary to reinvestigate the most suitable methods or techniques that will allow for handling dynamics and stochasticity of the grid [15]. Barriers to the smart grid development are developed based on the current existing grid power system structure.

6.2 BARRIERS AND SOLUTIONS TO SMART GRID DEVELOPMENT

Figure 6.1 illustrates the multilayered approach to designing the smart grid. The layers, or levels, [1, 2] are:

1. System Planning and Maintenance
2. Energy Auction

Smart Grid: Fundamentals of Design and Analysis, First Edition. James Momoh.
© 2012 Institute of Electrical and Electronics Engineers. Published 2012 by John Wiley & Sons, Inc.

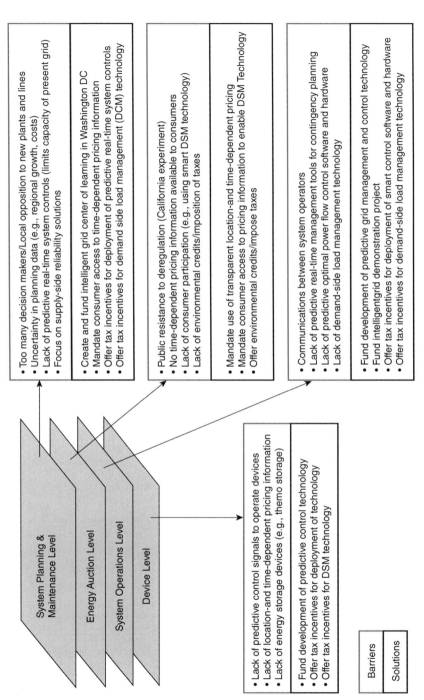

Figure 6.1. Barriers and solutions to the smart grid.

System Planning & Maintenance Level

Energy Auction Level

System Operations Level

Device Level

- Too many decision makers/Local opposition to new plants and lines
- Uncertainty in planning data (e.g., regional growth, costs)
- Lack of predictive real-time system controls (limits capacity of present grid)
- Focus on supply-side reliability solutions

- Create and fund intelligent grid center of learning in Washington DC
- Mandate consumer access to time-dependent pricing information
- Offer tax incentives for deployment of predictive real-time system controls
- Offer tax incentives for demand side load management (DCM) technology

- Public resistance to deregulation (California experiment)
- No time-dependent pricing information available to consumers
- Lack of consumer participation (e.g., using smart DSM technology)
- Lack of environmental credits/imposition of taxes

- Mandate use of transparent location-and time-dependent pricing
- Mandate consumer access to pricing information to enable DSM Technology
- Offer environmental credits/impose taxes

- Communications between system operators
- Lack of predictive real-time management tools for contingency planning
- Lack of predictive optimal power flow control software and hardware
- Lack of demand-side load management technology

- Fund development of predictive grid management and control technology
- Fund intelligentgrid demonstration project
- Offer tax incentives for deployment of smart control software and hardware
- Offer tax incentives for demand-side load management technology

- Lack of predictive control signals to operate devices
- Lack of location-and time-dependent pricing information
- Lack of energy storage devices (e.g., themo storage)

- Fund development of predictive control technology
- Offer tax incentives for deployment of technology
- Offer tax incentives for DSM technology

| Barriers |
| Solutions |

3. Systems Operations
4. Device

System Planning and Maintenance Level: Issues include too many decision-makers, local opposition to new plants and lines, planning uncertainties [3, 4], lack of predictive real-time system controls, and not enough focus on supply-side reliability solutions.

To overcome these barriers, the smart grid design needs consumer access to time-dependent pricing information, tax incentives for deployment of predictive real-time system controls, and tax incentives for DSM technologies. The solution method will:

- Create and fund an intelligent grid center of learning
- Mandate consumer access to time-dependent pricing information
- Offer tax incentives for deployment of predictive real-time system controls
- Offer tax incentives for DSM
- These solutions should minimize maintenance at the level of implementation.

Energy Auction Level: Issues include public resistance to deregulation, inadequate time-dependent pricing information unavailable to consumers, lack of consumer participation (for example, using smart DSM technology), and lack of environmental credits/imposition of taxes.

The solution will

- Mandate use of transparent location-and-time-dependent pricing
- Mandate consumer access to pricing information to enable DSM technology
- Offer environmental credits/impose taxes

Energy auction level (EAL) purports to redefine pricing structure for the energy market. This will go a long way toward making the accessibility and usage of the grid affordable by the general populace.

System Operation Level: Issues include communication pitfalls among system operators to handle emergencies, lack of predictive real-time management tools for efficient handling, and lack of predictive optimal power flow control software and hardware and DMS technology.

The solution will:

- Fund development of predictive grid management and control technology
- Fund intelligent demonstration projects
- Offer tax incentives for deployment of smart control software and hardware
- Offer tax incentives for DSM technology

Device Level: Issues include lack of predictive control signals to operate devices, lack of locational pricing and time-dependent pricing information, and lack of energy storage devices.

The solution will:

• Fund development of predictive control technology
• Offer tax incentives for deployment of technology
• Offer tax incentives for DSM technology

6.3 SOLUTION PATHWAYS FOR DESIGNING SMART GRID USING ADVANCED OPTIMIZATION AND CONTROL TECHNIQUES FOR SELECTION FUNCTIONS

Electric power grids are highly complex dynamical systems vulnerable to a number of disturbances in day-to-day operations. Random disturbances from weather and accidents rarely produce wide area [6, 10] catastrophic failures. However, the availability of detailed real-time operational data opens the door for maliciously designed disturbances. Figure 6.2 illustrates the four advanced optimization and control techniques required to meet the criteria for smart grid performance discussed in Chapter 3.

6.4 GENERAL LEVEL AUTOMATION

Automation at this level involves the use of addressed computation technologies and a new algorithm for dispatch and unit commitment to ensure:

• Reliability
• Stability
• Optimal dispatch
• Unit commitment under different uncertainties and constraints
• Security analysis
• Distributed generation control

Forecasting techniques must be incorporated into real-time operating practices as well as day-to-day operational planning, and consistent and accurate assessment of variable generation availability to serve peak demand is needed in longer-term system planning. High-quality and real-time data must be integrated into existing practices and software. The electricity industry is being encouraged to pursue research and development in these areas [1].

6.4.1 Reliability

Power system reliability is defined as the ability to deliver electricity to all points of power utilization within acceptable standards. The traditional reliability analysis methods are deterministic, for example, n-1 criterion where the system is deemed reliable if it can operate under a single unplanned outage [2], whereas in the smart grid

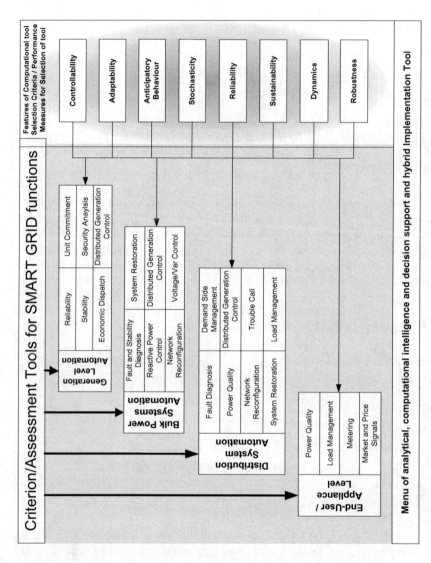

Figure 6.2. Smart grid using advanced optimization and control techniques.

environment, reliability analysis may not be used for extreme events. Thus, a new reliability and sustainability metric should be capable of handling:

- Nonlinearity of the system parameters, especially in the presence of the new mix of generation resources (wind, storage, hydro, and so on) and high power conversion devices
- Uncertainty in load demand and generation availability that are functions of time and contingencies
- System dynamics that reflect the market, availability of natural resources, network reconfiguration, and load switching
- Stochasticity of the system parameters due to man-made or natural events over different time-scales of operations

Indices such as loss of load probability (LOLP) and expected unserved energy (EUE) are used to determine the energy/load that can be supplied over time for different reliability services. The cost component balances the cost of reliability improvements with the cost of service disruptions. The ultimate goal is to achieve the maximum reliability under a variety of probable contingencies. Autonomous control actions will serve to meet this goal.

6.4.2 Stability

Power system stability is the ability of the bulk power electric power system to withstand sudden disturbances such as electric short circuits or unanticipated loss of system components [3]. At the generation level, transient stability is the primary consideration; as the system expands, the significance of transient stability increases.

In general, voltage and angle stability assessments are needed to ensure dynamic reconfiguration in response to system faults and disturbances. The methods discussed earlier in chapter 4 are conventional analytical methods including risk assessment and Lyapunov stability assessment. Since the margin of stability cannot be obtained using numerical methods, an energy-based method has been utilized for angle stability. Traditional methods for solving stability problems, including continuation power flow method, singular value decomposition, Jacobian condition number, and energy margin methods are generally used, but all are based on offline studies.

WAMS and PMU measurements prepared for the smart gird environment will allow for the real-time evaluation of the system under different loading and unknown contingencies. Advancements in stability studies aim to address the impacts of contingency with the help of developing new indices for stability assessment.

6.4.3 Economic Dispatch

Economic dispatch is a computational process where the total required generation including renewable energy resources is distributed among the generation units in operation by minimizing the selected cost criterion [4] subject to load and operational constraints. For any specified load condition, the output of each plant (and each generating unit within the plant) which will minimize the overall cost of fuel needed to serve the system load is calculated [3]. Traditionally, the economic dispatch problem is formulated as an optimization with cost as the quadratic objective function:

$$F(P_g) = \sum_{i=1}^{N_g} \left(\alpha_i + \beta_i P_{gi} + \gamma_i P_{gi}^2 \right)$$

$$\sum_{i=1}^{N_g} P_{gi} - \sum_{i=1}^{N_D} P_{D_i} + P_{loss} = 0$$

$$\text{s.t } V_{\min} \leq V \leq V_{\max}$$

$$P_{g\min} \leq P_g \leq P_{g\max}$$

The constraints include equalities and inequalities which represent the power equations as well as generator, bus, voltage, and line flow limits. This can be solved using analytical mathematical programming such as NLP, QP, LP, Newton method, interior point methods (IPM), and decision support methods such as AHP.

Computational methods used to solve this form of problem require simplifications of the objective function to be piecewise linear and monotonically increasing, for solution, but they suffer from the power system's problem of nonconvexity behavior which then leads to local minimization or local optimality. Alternative solution methods such as (EP) [4], GA [5], tabu search [6], neural networks [7], particle swarm optimization [8, 9], and ADP are proposed for improving the performance of the economic dispatch algorithm.

6.4.4   Unit Commitment

UC, an operation scheduling function, is sometimes termed "predispatch." In the overall hierarchy of generation resources management [10], the UC function coordinates economic dispatch and maintenance and production scheduling over time. UC scheduling covers the scope of hourly power system operation decisions with a one-day to one-week horizon.

The UC decisions are coupled or iteratively solved in conjunction with coordinating the use of hydro units including pumped storage with capabilities and to ensure system reliability. The function may also include labor constraints due to normal crew policy that a full operating crew will be available without committing to overtime costs. A prime consideration is the adoption of environmental controls, such as fuel switching. The application of computer-based UC programs by electric utilities has been slow due to the following reasons:

i.   They are not readily transferred between systems. The problem is so large and complex that only the most important features can be included; these vary among systems, thus requiring tailor-made applications.

ii.   Political problems, constraints, and peculiarities of systems that are not easily amenable to mathematical solutions and may be difficult to model.

iii.   Operating situation changes quickly; there is so much objective and subjective information about the system that the input requirements of sophisticated, computerized schedulers are discouraging.

iv.   As in other computer applications areas, developing fully workable systems has been difficult; building an operator's confidence is also difficult.

The UC schedule is obtained by considering:

* Unit operating constraints and costs
* Generation and reserve constraints
* Plant startup constraints
* Network constraints

Formulation of Unit Commitment. To account for constraints, randomness, and stochastic variables, several assumptions are needed for the formulation of UC. Several classifications for reserve include units on spinning reserve and units on cold reserve under the conditions of banked boiler or cold start part of the formulation. The first constraint to be met is that net generation must be greater than or equal to the sum of total system demand and required system reserve:

$$\sum_{i=1}^{N} P_{gi}(t) \geq (\text{Net Demand} + \text{Reserve})$$

In case the units should maintain a given amount of reserve, the upper bounds must be modified accordingly. Therefore:

$$P_{gi}^{max} = P_{gi}^{capacity} - P_{gi}^{reserve}$$

$$Demand + Losses \leq \sum_{i=1}^{N} P_{gi} - \sum_{i=1}^{N} P_{gi}^{reserve}$$

$$C_{cold} = C_O(1 - e^{\alpha t}) + C_L$$

where

C_{cold}: Cost to start an off-line boiler

α: Unit's thermal time constant;

t: Time in seconds

C_L: Labor cost to up the units

C_O: Cost to start up a cold boiler

$$C_{banked} = C_B t + C_L$$

where

C_B: Cost to start up a banked boiler

t: Time in seconds

Additional constraints include uncertainty and some of the security index or vulnerability index and congestion margin used in the formulation of the UC system for a smart grid.

Lagrangian relaxation is among the classical techniques used for the UC problem where the constraints are based on stochastic variables and predictivity is considered. The stochastic optimization method is useful for solving this problem as its randomness is a good solution approach, but its predictivity is not as good compared to ADP [13, 14]. ADP can handle discrete systems which use computational techniques and IPM as features of the more recent techniques which have been able to handle stochastic and dynamics indices.

6.4.5   Security Analysis

Power system security analysis detects whether the power system is in a secure state or an alert state. Secure state implies that the load is satisfied and no limit violations will occur under the current operating conditions and in the presence of unforeseen contingencies. Alert state implies that particular limits are violated and/or the load demand cannot be met, and corrective actions are necessary to bring the system back to the secure state [10].

Dynamic analysis evaluates the time-dependent transition from the precontingent to the postcontingent state. Dynamic security has been analyzed either by deriving dynamic security functions only, or along with the development of some preventive action techniques [11–13]. Traditionally, n-1 contingency analysis has been used for the evaluation of security defined as the system's ability to supply load under contingency.

The combination of computational intelligence and NP provides an improved security assessment which accounts for randomness or stochasticity in the data. The index for measuring security is based on real-time measurements. An example of the new indices to estimate system performance is vulnerability which could be based on loss of load index, voltage index, or power flow index.

6.5   BULK POWER SYSTEMS AUTOMATION OF THE SMART GRID AT TRANSMISSION LEVEL

The automation of different functions of the bulk grid level of the transmission system is important for achieving resilience and sustainability of the system. The following functions are evaluated, and the appropriate intelligent technology is proposed:

- Real-time angle, voltage stability, and collapse detection and prevention via intelligent-based data
- Reactive power control based on intelligent coordination controls
- Fault analysis and reconfiguration schemes based on intelligent switching operations

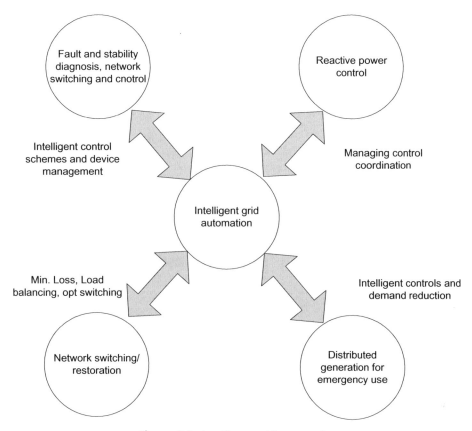

Figure 6.3. Intelligent grid automation.

- Power generation and load balance via intelligent switching operation and minimizing demand interruption
- DG and DSM via DR strategy for peak shaving, including increased proliferation and control of RER

The integration of these functions is shown in Figure 6.3.

Some of the functions and roles are presented next.

6.5.1 Fault and Stability Diagnosis

Voltage and angle stability assessments ensure dynamic reconfiguration in response to system faults and disturbances. Since the margin of stability cannot be obtained using numerical methods, an energy-based method has been proposed for angle stability. To date, real-time evaluation of the system under different loading and unknown contingencies has not been implemented. Analytical techniques used for fault analysis include the circuit theory-based method and the traveling wave-based method. Methods for detection of voltage instability are generally based solely on local measurements of voltage, for example under voltage load shedding (UVLS). WAM and PMUs also play a major role in stability monitoring and analysis.

6.5.2 Reactive Power Control

The main factor causing instability is the inability of the power system to meet the demand for reactive power. The heart of the problem is usually the voltage drop that occurs when active power and reactive power flow through the inductive reactance associated with the transmission network [14]. Reactive power control uses optimization techniques to minimize loss and schedule reactive resources to remove the voltage degradation problem. The use of real-time power margin as a constraint in optimization can be solved in real time with intelligent optimization techniques based on DSOPF.

Other functions such as network switching and restoration use knowledge based systems completed with GA and fuzzy logic to select the outage units and lines and appropriate control measures. for scheduling outages. ADP and other variants of EP, such as GA and ACO, are adapted to address RER penetration into the smart grid and to manage variability.

6.6   DISTRIBUTION SYSTEM AUTOMATION REQUIREMENT OF THE POWER GRID

Distribution automation and control involves the delivery of energy to customers. Figure 6.4 shows the distribution automation schemes for distribution systems.

Smart Distribution solutions are designed to minimize energy losses, mitigate power disruptions, and optimally utilize distributed smart grid components, including alternative energy sources, power storage, and PHEV charging infrastructure.

Greater access to larger pools of available generation and demand may also be important to the reliable integration of large-scale variable generation. As the level of variable generation increases in a balancing area, the resulting variability may not be manageable with its existing conventional generation resources. Base load generation [15] may need to be frequently cycled in response to these conditions, which introduces reliability concerns as well as economic consequences. These efforts may also help to address minimum load requirements of conventional generation and contribute to the effective use of off-peak, energy-limited resources.

6.6.1   Voltage/VAr Control

Voltage control within a specified range of limits and capacitor switching is an effective means of minimizing loss, improving voltage profiles, and deferring construction in the end within the reliability and power quality constraints.

Voltage/VAr Control considers a multiphase unbalanced distribution system operation, dispersed generation, and control equipment in the large system. In distribution automation, functions using voltage/VAr control options must maintain proper com-

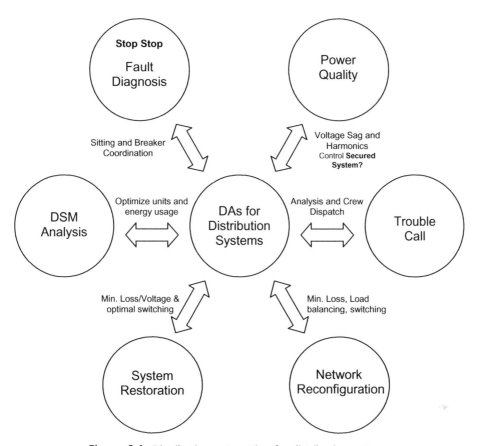

Figure 6.4. Distribution automation for distribution systems.

munication throughout decision-making to install capacitors, thereby realizing the cost benefit.

Formulation of Voltage/VAr Problem. This includes integrating voltage/VAr with the load management problem to improve efficiency. The objectives are:

a. Customer Outage Cost

$$\text{Min Outage Cost} = \sum_{i} \left(X_{ik} \cdot PC_{ki} \cdot CC_{k} \right)$$

where

X_{ik}: Level of Curtailable Load Selection of type k at bus i (p.u. MW)
PC_{ki}: Max Curtailable MW of type k at the ith bus (p.u. MW)
CC_{k}: Curtailment Cost of Customer Type k ($/p.u. MW)

b. Loss Minimization

The objective of the loss minimization function is given by:

$$\text{Min } I^2 r = \sum_{ij} r_{ij} \frac{P_{ij}^2 + Q_{ij}^2}{|V_i|^2}$$

c. Load Balancing

$$Max \frac{S_i}{S_{i\max}}$$

where:

N: Total number of buses
P_{ij}, Q_{ij}: Transfer Power (branch i-j at p.u.)
V_i: Voltage of bus i (p.u.)

d. Multiple objective functions

$$Z = Min[a + b - c]$$

System Operating Constraints include:

a. Branch Flow Equations

$$Y_{i+1} = f_{i+1}(Y_i)$$

where $Y_i = [P, Q, |V|^2, X_k, Q_s, \delta_i]$
Branch flow considers the recursive relationships between the successive nodes in the radial distribution system. The demand in P_D, Q_D has an interruptible component which is the load management control options X. In addition, the reactive power equation has the capacity switching option Q_s.

b. Voltage Limits/Current Limits

The voltage and current limits are given as:

$$|V_i^{\min}| \le |V_i| \le |V_i^{\max}|$$

$$|I_{ij}^{\min}| \le |I_{ij}| \le |I_{ij}^{\max}|$$

c. Capacitor Control Limits

$$Qs_i^{\min} < Qs_i \le Qs_i^{\max}$$

d. Curtailable Load Control Limits

$$P_i \times X_{ki} \leq Pc_{ki}$$
$$Q_i \times X_{ki} \leq Qc_{ki}$$

6.6.2 Power Quality

Power quality is the ability of a system to operate without causing disturbance or damage to loads and components. Power quality is a major concern because of the sensitivity of digital and modern control equipment to distortion/PQ deterioration. A basic requirement for maintaining power quality is balancing supply and demand. Typical indices utilized as power quality measures include:

- Voltage transient impulse
- Voltage sag
- Total harmonic distortion
- Flicker factor

A complete listing of quality measures employed in the global electric power industry is given in Table 6.1.

Several accepted power quality standards which apply to electrical distribution systems are itemized in "Voltage Disturbances Standard EN 50160: Voltage Characteristics in Public Distribution Systems." A real-time power quality study features several aspects [13]:

TABLE 6.1. Typical Power Quality Indices

Phenomena Indices	Calculations	Parameters	Networks/Systems where index is applied
Voltage Sag or Swell	$\Delta V = |V_i - V_{nom}|$	V_i—voltage at bus i V_{nom}—nominal voltage	Networks which feature: Remote System Faults
Flicker factor	$FF = \dfrac{\Delta V}{|V|}$	V—voltage	Networks which feature: Intermittent loads Motor starting Arc Furnaces
Total Harmonic Distortion (THD)	$THD = \dfrac{\sqrt{\sum\limits_{i=2}^{\infty} \left(I^{(i)}\right)^2}}{I^{(1)}}$ $THD = \dfrac{\sqrt{\sum\limits_{i=2}^{\infty} \left(V^{(i)}\right)^2}}{V^{(1)}}$	$I^{(i)}$—i^{th} harmonic current $V^{(i)}$—i^{th} harmonic voltage	Networks which feature: Non-linear loads System resonance

1. Sensitive detection of power disturbances
2. Real-time measurement of parameters of signal components in power disturbances
3. Quantification of power disturbances and their negative impacts on power systems
4. Identification of types and causes of power disturbances
5. Location of power disturbances

Power quality monitoring and analysis in the smart grid will utilize real-time data.

6.6.3 Network Reconfiguration

Cost-effective reconfiguration assessment strategies are needed to address the following challenges:

- Nonlinearity of power system parameters
- Uncertainty in load demand and generation availability
- System dynamics
- Stochastic of system parameters

Intelligent control methods including adaptive critics designs, fuzzy logic, and ANN methods may provide adaptive stochastic solutions to the general problem formulation.

6.6.4 Demand-Side Management

DSM is an effective means of modifying customer demand to reduce operating expenses from expensive generators and defer capacity addition in the long run. Environmental conservation is improved with DMS options, which also sustain industrialization at minimum cost. In addition DSM options provide reduced emission reduction in fuel production and reduced costs, and contribute to system reliability. DSM options are categorized into:

- Peak shifting
- Valley filling
- Peak clipping
- Storage conservation

These options have an overall impact on the utility load curve. For distribution automation functions, DSM is classified into three categories:

1. Direct Control of Load [12]: Uses a communication system such as a power line carrier/radio to transmit direct control of load, small generators, and storage from utility to customers.

2. Local Load Control Option: Enables customers to self-adjust loads to limit peak demand, for example, demand activated breakers, load interlocks, timers, thermostats, occupancy sensors, cogeneration heating, cooling storage, and so on.

3. Distribution Load Control: Utility controls customer loads by sending real-time prices.

6.6.5 Distribution Generation Control

More comprehensive planning approaches from the distribution system to the bulk power system are needed, including probabilistic approaches at the bulk system level. This is particularly important with the increased prevalence of distributed variable generation, for example, local wind plants and rooftop solar, on distribution systems. In aggregate, distributed variable generation needs to be treated, where appropriate, similarly to transmission-connected variable generation. The issues of note include forecasting, restoration, voltage ride through, safety, reactive power, observability, and controllability [11, 14]. Standard, nonconfidential, and nonproprietary power flow and stability models are needed to support improved planning efforts and appropriately account for new variable resources. Variable generation manufacturers are encouraged to support the development of these models.

6.7 END USER/APPLIANCE LEVEL OF THE SMART GRID

At the end-user level, significant changes in metering and monitoring are being introduced. DSM and DR are the two fundamental automation functions which will be enhanced by developments at the end-user and appliance levels. DSM includes customers as utility planning options. The total system cost including DSM cost is minimized to obtain an optimal mix of the supply side of the generation and demand side load reduction. Analysis of DSM uses techniques such as daily load curve or mathematical programming methods. DSM has been carried out using the context of unit commitment studies, optimal power flow studies, load reduction forecasting methods, engineering features of the end-user equipment interruptible load management program, survey methods (data collection), and DP to optimize energy procurement and load management by utilities.

Standards for integrating advanced mobile power recovery units, energy storage, and on-demand or peak-shaving technologies will invariably necessitate the development of standards for coordination, communication, and interoperability of multiple appliances competing for energy at the optimum grid price available. These and other related issues are research topics that require urgent attention.

6.8 APPLICATIONS FOR ADAPTIVE CONTROL AND OPTIMIZATION

There are several potential applications for adaptive controls and optimization in the development of the smart grid's design framework. It is important that adaptation and control strategy scheduling consider the stochastic and dynamically interdependent attributes of power systems. Conventional techniques to handle these problems often suffer from the curse of dimensionality and somewhat heuristic methods which tend to give nonoptimal solutions. Therefore, system stochasticity and dynamic interdependence lead to unstable, nonfeasible, and nonoptimal solutions.

In DSOPF, the efficient optimization technique typically based on two-stage action and critic networks are used to achieve:

- Multiobjective, time-dependent optimization for complex systems
- Optimal scheduling subject to technical constraints of the plant or system
- Adaptation to perturbation of power system dynamics over time
- Adaptation to varying random noise, uncertainties, and/or missing or corrupted measurements
- Adaptation to changes in system structure while distinguishing between observable and unobservable measurements

6.9 SUMMARY

This chapter has discussed the development of predictive grid management and control technology for the deployment of smart control software and hardware to enhance smart grid performance. This will lead to benefits for customers and stockholders [14, 15]. The lack of predictive control signals to operate devices and the lack of energy storage devices were described. Research on the integration of renewable and storage and new software and control technology was suggested to deploy, operate, and maintain the new grid's infrastructure.

REFERENCES

[1] "Accommodating High Levels of Variable Generation," North American Electric Reliability Corporation, 2009.

[2] J.A. Momoh and M.E. El-Hawary. *Electric Systems, Dynamics and Stability with Artificial Intelligence Applications*. New York: Marcel Dekker, 2000.

[3] A.J. Wood and B.F. Wollenberg. *Power Generation, Operation and Control*. New York: John Wiley & Sons, 1996.

[4] H. Yang, P. Yang, and C. Huang. "Evolutionary Programming Based Economic Dispatch with Non-Smooth Fuel Cost Functions," *IEEE Transactions on Power Systems* 1996, 11, 112–118.

[5] D.C. Walters and G.B. Sheble. "Genetic Algorithm Solution of Economic Dispatch With The Valve-point Loading," *IEEE Transactions on Power Systems* 1993, 8, 1325–1332.

[6] W.M. Lin, F.S. Cheng, and M.T. Tsay. "An Improved Tabu Search For Economic Dispatch With Multiple Minima," *IEEE Transactions on Power Systems* 2002, 17, 108–112.

[7] K.Y. Lee, A. Sode-Yome, and J.H. Park. "Adaptive Hopfield Neural Network For Economic Load Dispatch," *IEEE Transactions on Power Systems* 1998, 13, 519–526.

[8] R.C. Eberhart and Y. Shi. "Comparing Inertia Weights and Constriction Factors in Particle Swarm Optimization," Proceedings of the 2000 Congress on Evolutionary Computation, 1, 84–88.

[9] R.C. Eberhart and Y. Shi. "Particle Swarm Optimization: Developments, Applications, and Resources," Proceedings of the 2001 Congress on Evolutionary Computation, 1, 81–86.

[10] OSI. Smart Grid Initiatives White Paper: Revision 1.1.

[11] A.K. Al-Othman, F.S. Al-Fares, and K. M. EL-Nagger. "Power System Security Constrained Economic Dispatch Using Real Coded Quantum Inspired Evolution Algorithm," *World Academy of Science, Engineering and Technology* 2007, 29, 7–14.

[12] R.B. Alder. "Security Consideration Economic Dispatch With Participation Factors," *IEEE Transactions on Power Apparatus and Systems* 1977, PAS-96.

[13] M. El-Sharkawy and D. Niebur. "Artificial Neural Networks With Application to Power Systems," IEEE Power Engineering Society, A Tutorial Course, 1996.

[14] F.F. Wu, J.H. Chow, and J.A. Momoh, Eds. *Applied Mathematics for Restructured Electric Power Systems: Optimization, Control, and Computational Intelligence.* Springer Science and Business Media, Inc., 2005: pp. 11–24.

[15] T. Lin and A. Domijan. "On Power Quality Indices and Real Time Measurements," *IEEE Transactions on Power Delivery* 2005, 20, 2552–2562.

SUGGESTED READING

P. Kundur. *Power System Stability and Control.* Washington, D.C.: McGraw-Hill Inc., 1994.

7

RENEWABLE ENERGY AND STORAGE

7.1 RENEWABLE ENERGY RESOURCES

The design and development of the smart grid requires modeling renewable energy sources and technologies such as wind, PV, solar, biomass, and fuel cells, analyzing their levels of penetration, and conducting impact assessments of the legacy system for the purpose of modernization. The roadmap envisions widespread deployment of Distributed Energy Resources (DERs) in the near future. Renewable technologies have been positioned to reduce both America's dependence on foreign oil and the environmental impacts of energy production [1]. Renewable energy technologies and their integration introduce several issues including enhancement of efficiency and reliability, and the development of state-of-the-art tracking to manage variability.

Architecture designs which include optimal interconnections, optimal sizing and siting DERs for optimum reliability, security, and economic benefits are also critical aspects. Additionally, computational development of the smart grid to permit estimation and forecasting models for fast real-time accurate predictions of these variable power sources need to be addressed. The challenges and the technologies including conversion, storage, and PHEV, are discussed in this chapter.

Smart Grid: Fundamentals of Design and Analysis, First Edition. James Momoh.
© 2012 Institute of Electrical and Electronics Engineers. Published 2012 by John Wiley & Sons, Inc.

7.2 SUSTAINABLE ENERGY OPTIONS FOR THE SMART GRID

Sustainable energy is derived from natural sources that replenish themselves over of time. Sometimes called green power, because they are considered environmentally friendly and socially acceptable, they include sun, wind, hydro, biomass, and geo-thermal.

The future electric grid [3, 4] will feature rapid integration of alternative forms for energy generation as a national priority. This will require new optimization for energy resources that are distributed with interconnection standards and operational constraints. The legacy grid was designed to handle this change. The stability, reliability, and cost implications of renewable energy resources in the development of technical and economic software are vital.

Renewable energy options are meant to provide the smart grid with:

i. Remote utilization and storage of RER resources output

ii. Enhancement of functionality of grid-connected renewable energy systems (RES)

    a. Facilitating give-and-take of energy from the system

    b. Redistribution/reallocation of unused power from grid-connected RES

    c. Facilitating storage of grid-generated and RER-generated energy by back-up storage technologies at customer end

    d. Tracking interactions for billing and study

iii. Enhancement of functionality of electric vehicles and plug-in hybrids

iv. Utilization of vehicle battery packs as energy storage devices

The common RER uses in smart grid networks are presented below.

7.2.1 Solar Energy

Solar energy harnessed by the use of photovoltaic (PV) cells was first discovered in 1839 by French physicist Edmund Becquerel. The technology can be a single panel, a string of PV panels, or a multitude of parallel strings of PV panels. Solar PV has no emissions, is reliable, and requires minimum maintenance.

The PV system generally considers:

i. Insolation: The availability of solar energy conversion to electricity. Insolation levels are affected by the operating temperature of PV cells intensity of light (location-dependent), and the position of the solar panels (maximize the power tracking while maximizing perpendicular incident light rays).

ii. Emission: PV emission levels are environmental friendly.

To improve efficiency, the materials used for manufacturing PV cells include amorphous silicon, polycrystalline silicon, cadmium telluride, microcrystalline silicon, and copper indium selenide.

7.2.2   Solar Power Technology

Solar power technology enhances PV output by concentrating a large area of sunlight into a small beam using lenses, mirrors, and tracking systems. Parabolic troughs and solar power towers are examples of such technologies [12].

Cost Implication: Manufacturers continue to reduce the cost of installation as new technology is developed for manufacturing materials. Much work is being conducted in the manufacturing of PV and the development of superior materials. Basic materials include monocrystalline and polycrystalline panels, cast polycrystalline silicon, and string and ribbon silicon, as well as amorphous silicon or thin film panels which are created by the application of amorphous silicon, copper indium diselenide (CIS), and cadmium telluride (CdTe) as a thin semiconductor film. This application allows for the manufacturing of panels that are less time-consuming to manufacture with lower manufacturing costs and that are applicable to varied applications, albeit with reduced efficiency.

7.2.3   Modeling PV Systems

There are several PV simulation programs which allow for series analysis for time such as PV-DesignPro [5]. The input typically consists of one year's worth of hourly global-direct irradiance, and temperature and wind speed data, for example, TMY data, array geometry, and other PV and balance of system (BOS) parameters. Many models exist for the calculation of the power output of a PV cell or bank. Due to the varying efficiencies and numerous technologies presently available, power output is affected by environmental conditions and module specifications. The I-V characteristic model of a single cell is commonly used for PV technologies. The model is given by:

$$I = I_{sc} - I_{os}\left(e^{\frac{q(V+IR_s)}{nkT}} - 1\right) - \frac{V + IR_s}{R_{sh}}$$

$$I_{os} = AT^{\gamma}e^{\left(\frac{-E_g}{nkT}\right)}$$

$$I_d = I_{sc} - I_{os}\left(e^{\left(\frac{q}{nkT}E_d\right)} - 1\right)$$

where for an array of $N_s \times N_{sh}$ solar cells:

$$I_d = \frac{I_{d\_mod}}{N_{sh}}$$

$$E_{d\_mod} = E_d \times N_s$$

$$R_{s\_mod} = R_s\left(\frac{N_s}{N_{sh}}\right)$$

$$R_{sh\_mod} = R_{sh}\left(\frac{N_s}{N_{sh}}\right)$$

where

I: current flowing into load of a solar cell (A)

I_{sc}: short circuit current (A)

I_{os}: saturation current (A)

s: insolation (kW/m^2)

q: electron charge (1.6×10^{-19} (C))

k: Boltzmann constant (1.38×10^{-23} (JK^{-1}))

T: p-n junction temperature (K), t (°C)

N: junction constant

A: temperature constant

γ: temperature dependency exponent

E_g: energy gap (eV)

V: voltage across solar cell (V)

V_{oc}: open circuit voltage of a solar cell (V)

R_s, R_{s_mod}: series parasitic resistance for cell and entire module (Ω)

R_{sh}, R_{sh_mod}: shunt parasitic resistance for cell and entire module (Ω)

E_d, E_{d_mod}: across voltage of an ideal solar cell and entire module (V)

I_d, I_{d_mod}: current of an ideal solar cell and entire module (A)

N_s: number of series cell junctions of a PV module

N_{sh}: number of parallel cell junctions of a PV module

V_{out}: across voltage of a PV module (V)

I_{out}: current of a PV module (A)

R: connected load (Ω)

An alternative equation for the modeling of the output of the PV panels is [6]:

$$P_{mp} = \frac{G}{G_{ref}} P_{mp,ref}\left[1 + \gamma\left(T - T_{ref}\right)\right]$$

where

G is the incident irradiance

P_{mp} is the maximum power output

$P_{mp,ref}$ is the maximum power output under standard testing conditions

T is the temperature

T_{ref} is the temperature for standard testing conditions reference (25°C)

$G_{ref} = 1000$ Wm^{-2}

γ is the maximum power correction for temperature

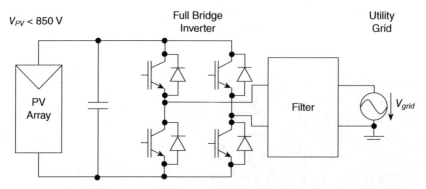

Figure 7.1. PV inverter system for DC-AC conversion.

The PV output is variable due to in relation to the solar insolation and surface temperature. The data for predicting the solar input is several years of measurements of irradiance on the proposed locator. These statistical measures may be estimated from meteorological data available from the site, from a nearby site having similar irradiance, or from an official solar atlas or database [7]. Solar insolation has been modeled probabilistically for variability studies of PV systems utilizing various distributions including Gaussian (normal) and Beta probability density functions.

Conversion and Power Electronic Technology. Several inverter systems convert or transform the DC into AC for grid-connected PV systems (see Fig. 7.1).

Penetration of PV into the smart grid requires studies in variability and conversion technology. The mathematical models and probability density used to model PV behavior include Beta and Rayleigh density functions. The tracking method based on fuzzy/ GA technologies is able to obtain enough power points for effective delivery. Siting and sizing of PV uses classical and computational intelligence methods that involve decisions based on real-time data.

7.2.4 Wind Turbine Systems

Wind is one of the fastest-growing sources of renewable energy throughout the world (see Fig. 7.2). Turbines produce electricity at affordable cost without additional investments in infrastructure such as transmission lines. A wind turbine consists of a rotor, generator, blades, and a driver or coupling device. Compared with PV, wind is the most economically competitive renewable.

Table 7.1 shows various wind turbine configurations. Both wind speed and the height of pole-mounted units contribute to the power output.

Although turbines produce no CO_2 or pollutants [9], wind has three drawbacks: output cannot be controlled, wind farms are most suited for peaking applications, and power is produced only when there is sufficient wind.

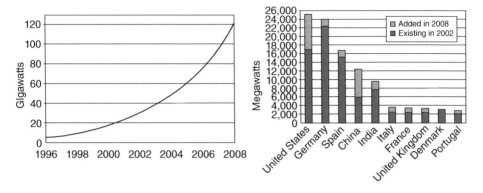

Figure 7.2. (a) Wind power, existing world capacity, 1996–2008; (b) wind power capacity, top ten countries, 2008.

Modeling Wind Turbines: Machines which consume reactive power and produce real power. The quantification of the capacity/real power output is given by:

$$P_m = \frac{1}{2}\rho \cdot \pi R^2 \cdot V^3 \cdot C_p$$

where

ρ is the air density (kg/m^3)

R is the turbine radius (m)

C_p is the turbine power coefficient power conversion efficiency of a wind turbine

V is the wind speed (m/s)

The electrical power output is given by:

$$P_e = n_o P_m$$

where

$n_o = \eta_m \eta_g$

η_m, and η_g are the efficiency of the turbine and the generator, respectively

Wind energy options are typically random and require a modeling technique based on probability such as Gaussian, Rayleigh, and Beta pdfs. Another challenge is congestion management. Due to time-varying behavior and randomness, EP such as GA, ANN, and fuzzy logic and its hybrids are used to allocate wind energy to meet in-time local demand. The process includes obtaining real-time measurement, and using the system state, location, and network limitation to define new algorithms.

7.2.5 Biomass-Bioenergy

Bioenergy is the energy derived from organic matter such as corn, wheat, soybeans, wood, and residues that can produce chemicals and materials. Biopower is obtained

TABLE 7.1. Wind Turbine Modeling

Type	Type A: Constant Speed Wind Turbine	Type B : Variable Speed WT with variable rotor resistance	Type C : Variable Speed WT with partial-scale frequency converter	Type D : Variable Speed WT with full-scale frequency converter ()
Configuration				
Generator type	Squirrel Cage Induction Generator (SCIG)	Wound rotor induction generator (WRIG)	Doubly-fed induction generator (DFIG)	Electrically excited (wound rotor synchronous generator WRSG) or permanent magnet excited type (permanent magnet synchronous generator PMSG).
Reactive power	Draws reactive power from grid; must be accompanied by a capacitor back.	A capacitor bank performs the reactive power compensation and smooth grid connection occurs by means of a soft starter.	Converter performs the reactive power compensation and a smooth grid connection	Converter performs the reactive power compensation and a smooth grid connection
Key characteristics	Wind fluctuations are converted into mechanical fluctuations and further into electrical power fluctuations Does not support any speed control requires a stiff grid, mechanical construction must be able to support high mechanical stress caused by wind gusts. Limited active power	Total rotor resistance is controllable by variable additional rotor resistance, which is changed by an optically controlled converter mounted on the rotor shaft		Full variable speed controlled wind turbine, with the generator connected to the grid through a full-scale frequency converter

146

from a process called gasification, converting gas to gas turbines to generate electricity. Biomass can be converted directly into fluid fuels such as ethanol, alcohol or biodiesel derived from corn ethanol.

Biomass plants are commercially available in the United States for up to 11GW of installed capacity. Biomass power ranges from 0.5 GW to 3.0 GW using landfill gas and forest products, respectively. Biomass has traditionally been used for domestic cooking and heating in developing countries. It can produce power only when sufficient bioproducts are available and the conversion process is undertaken. Biomass produces CO_2 and other emissions. The desirable scheduling and allocation strategy for biomass in real time requires the capability to include variability of the modeling using new system theory concepts.

7.2.6 Small and Micro Hydropower

Hydropower is by far the largest renewable source of power/energy [8]. Small hydropower systems vary from 100 kW to 30 MW while micro hydropower plants are smaller than 100 kW. Small hydropower generators work in variable speed because of water flow. Induction motors provide a generator for a turbine system. The hydraulic turbine converts the water energy to mechanical rotational energy.

Small and micro hydropower systems are RER optimizations to enhance the smart grid. The issues of reliability and modeling are addressed as in PV and wind energy. CI technology is used for performance study and commitment.

7.2.7 Fuel Cell

Fuel cells can also be used to enhance power delivery in the smart grid. They are simply fuels from hydrogen, natural gas, methanol, and gasoline. The efficiency for fuel to electricity can be high as 65% since it does not depend on Carnot limits. Fuel cells are environmentally friendly by efficient use of fuel. Different forms include phosphoric acid (PAFC), proton exchange membrane (PEM), solid polymer molten carbonate (MCFC), solid oxide (SOFC), alkaline (a direct methanol), regenerative fuel cells, and botanic ceramic fuel cells (PCFC). Fuel cells produce virtually no emissions. Their cost is significantly high compared with conventional technologies.

The topology of a fuel cell is a stack which consists of the part of the fuel cell that holds the electrodes and electrolytic material. Hydrogen is extracted from gasoline propane, with natural gas refineries to operate fuel cells commercially.

Emission is very low and so these cells have minimum environmental impacts. The high efficiency leads to low fuel costs and minimum maintenance due to lack of moving parts. They have virtually no pollutant emission as CO_2 is rather low. They are a good fit for green power and premium power due to their quality. In addition, they provide a moderate high thermal quantity output and hence are ideal for CHP. However, they perform poorly as peaking power due to extremely high capital costs.

The efficiency of fuel cells ranges from 40–80%. Two commonly used fuel cell types are phosphoric acid fuel cells (PAFC), which operate at relatively high temperatures and use an external water cooling system to cool the stack, and proton exchange membrane fuel cells, which operate at a lower temperature than most other fuel cells and contain no chemicals such as liquid acids or molten bases that would cause concerns about materials of construction.

7.2.8   Geothermal Heat Pumps

This form of power is based on accessing the underground steam or hot water from wells drilled several miles into the earth. Conversion occurs by pumping hot water to drive conventional steam turbines which drive the electrical generator that produces the power. The water is then recycled back into earth to recharge the reservoir for a continuous energy cycle. There are several types of geothermal power plants including dry steam, flash stream, and binary cycle. Dry steam plants draw water from the reservoirs of steam, while both flash steam and binary cycle plants draw their energy from the recycled hot water reservoir.

In the United States, geothermal power is produced in California, Utah, Nevada, and Hawaii. Various applications of geothermal power include heat pumps, agriculture, fishing, farming, and food processing. Geothermal projects require significant upfront capital investment for exploration, drilling wells, and equipment. Exploration risk and environmental impacts are considered in geothermal power plant projects.

7.3   PENETRATION AND VARIABILITY ISSUES ASSOCIATED WITH SUSTAINABLE ENERGY TECHNOLOGY

The degree of penetration by sustainable energy into today's grid varies by location, depending on the availability of the conditions necessary for viable utility. This can be quantified by:

$$\text{Penetration Level} = \frac{\sum_{\forall i} P_{DG,i}}{\sum_{\forall j} P_{demand,j}}$$

The selection and implementation of the available sustainable energy technologies are subject to the issues of variability associated with the sources, for example, solar and wind. For solar energy, the variation of the solar radiation is a function of the time, location, height above the earth's surface, local weather conditions, and degrees of shading. The variability of the PV source is a function of the solar insolation which has been modeled using a Beta distribution function. Utilizing historical data for location and time allows calculation of parameters of the Beta model by solving equations given as

$$\alpha = \mu \left(\frac{(1-\mu)\mu}{\sigma} - 1 \right)$$

$$\beta = (1-\mu) \left(\frac{(1-\mu)\mu}{\sigma} - 1 \right)$$

The corresponding probability distribution function is formulated as $f(s)$:

$$f(s_i) = \frac{s_i^{\alpha-1}(1-s_i)^{\beta-1}}{\Gamma(\alpha)\Gamma(\beta)}\Gamma(\alpha+\beta)$$

There are three apparent sources of variability in wind technology: wind speed random variability, turning off turbines to prevent damage when wind speed values are greater than the tolerable limit, and no output when wind speed is less than the cut-in speed. The Weibull model, which has been used for studying wind speeds since 1978, models wind speed trends in numerous instances. The probability density function is:

$$f(V) = \begin{cases} \dfrac{k}{\lambda}\left(\dfrac{V}{\lambda}\right)^{k-1}\exp\left(\dfrac{V}{\lambda}\right)^k, & V \geq 0 \\ 0, & V < 0 \end{cases}$$

with mean and variance calculated in terms of the shape and scale parameters

$$\mu = \lambda\Gamma\left(1+\frac{1}{k}\right)$$

$$Var = \lambda^2\Gamma\left(1+\frac{2}{k}\right) - \mu^2$$

Utilizing historical data for location and time, the parameters of the Weibull model can be determined by simultaneously solving the mean and variance equations given as

$$k = \left(\frac{\sigma}{\bar{V}}\right)^{-1.086}$$

$$\lambda = \frac{\bar{V}}{\Gamma\left(1+\left(\dfrac{1}{k}\right)\right)}$$

Another common model is the Rayleigh distribution function, a special case of the Weibull model, which offers the added simplicity of only one parameter. This assumption can be made in several instances where the wind scale parameter in the location under study is approximately [2].

The concept of smoothing mitigates the impact of variability by utilizing several turbines over an area. From this penetration level an optimal scheme based on the computational intelligence technique is proposed for evaluating system and stability assessment in real time.

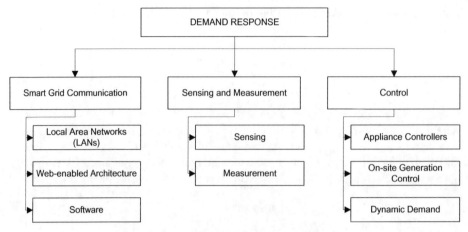

Figure 7.3. Demand Response technology tree.

7.4 DEMAND RESPONSE ISSUES

Energy management involves controlling electrical and mechanical systems to reduce power needs and the associated costs. DR helps to reduce customer demand on the grid that is dependent on that demand. Abundant data, including price signals and grid conditions, allow DR technologies to play a significant role. Monitoring operating parameters such as voltage, angle, and frequency of the system are utilized through real-time sensors in addition to controllers, metering signals, and two-way digital communication, in responding to changes in the grid and electricity prices. Automatic DR in times of disruption is a key feature of the smart grid, as are smart meters, smart appliances, and distributed RER. To achieve optimal control of demand and fulfill economic and environmental goals, utilities can show customers how to adjust their consumption to off-peak time demand to assist in efficient supply.

Figure 7.3 shows the DR applications that can be categorized into four components:

A. Energy Efficiency
B. Price-based DR
    a. Time-of-use (TOU)
    b. Day-ahead hourly pricing
C. Incentive-based DR
    a. Capacity/ancilliary services
    b. Demand bidding buy-back
    c. Direct load control
D. Time scale commitments and dispatch
    a. Years of system planning
    b. Months of operational planning

    c. Day-ahead economic scheduling

    d. Day-of economic dispatch

    e. Minutes dispatch

7.5 ELECTRIC VEHICLES AND PLUG-IN HYBRIDS

The integration of electric vehicles and hybrids is another component of the smart grid system. Vehicle-to-grid power (V2G) uses electric-drive vehicles (battery, fuel cell, or hybrid) to provide power for specific electric markets. V2G can provide storage for renewable energy generation and stabilize large-scale wind generation via regulation. Plug-in hybrids can dramatically cut local air pollution. Hybridization of electric vehicles and connections to the grid overcome limitations of their use including cost, battery size/weight, and short range of application. PHEVs provide the means to replace the use of petroleum-based energy sources with a mix of energy resources (encountered in typical electric power systems) and to reduce overall emissions [9].

    PHEVs offer advantages and disadvantages compared to other proposed solutions. The deployment of PHEVs potentially has a substantial positive impact on the electric power system from the point of view of increasing electric energy consumption, offsetting petroleum fuels with alternative sources of energy.

7.6 PHEV TECHNOLOGY

Through V2G power, a parked vehicle can provide power to the grid as a battery-electric, fuel-cell, or even a plug-in hybrid vehicle. Stored or unused energy that utilities reserve at night or during off-peak times can be used to support the vehicles in cases of extreme emergencies, for example, a significant decrease in oil supplies or a sudden rise in oil prices. It should be noted that electric drive systems are considered 70% efficient, for example, a first-generation plug-in hybrid can travel approximately 3–4 miles per kWh, or about 75 cents per gallon of gas.

    Each PHEV vehicle will be equipped with a connection to the grid for electrical energy flow, a control or logical connection necessary for communication with the grid operator, and onboard controls and metering. Figure 7.4 schematically illustrates the connections between vehicles and the grid from Reference 11. Electricity flows throughout the grid from generators to end-users while unused energy flow backs from the EDVs as shown by the lines with two arrows in Figure 7.4 Note that during times of low demand, battery EDVs can charge and discharge as necessary for power supply or demand. Fuel cells can create power from liquid and gaseous fuels and plug-in hybrids can function in either capacity [11].

7.6.1 Impact of PHEV on the Grid

PHEVs are expected to take 10% of the U.S. market share of new vehicles sold in 2015, and 50% of the market share by 2025 [8]. Utilities have become concerned with the

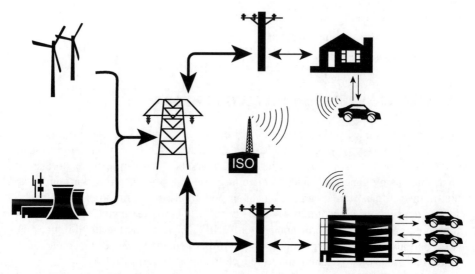

Figure 7.4. Schematic of proposed power line and wireless control connections between electric vehicles and the grid.

number of PHEVs coming on the market because there may be insufficient supply for the increased demand resulting from additional load for battery charging. By 2040, the addition of PHEV battery charging in the United States will increase existing load by 18% [8]. Unfortunately, this increase in load will eventually cause voltage collapse in amounts up to 96% of the nominal voltage in some areas, requiring the integration of transformers, capacitors, and other power distribution devices for mitigation.

It will be critical to study the trends of daily PHEV power usage and the average power consumption over one day to determine the impacts on the grid, market, environment, and economy. Peak driving levels will coincide with existing early and end-of-day peaking hours when users are waking up and heading to work and when they are home, having supper, and so on. During peak hours the increased need for energy may require users to charge their PHEVs in high-peak hours. Regulation of charge times will help maintain stability throughout the network.

7.7 ENVIRONMENTAL IMPLICATIONS

Environmental impact mitigation is a major driver of smart grid development. The integration/facilitation of the use of renewable energy resources for generation and the move towards the use of PHEVs are two critical aspects of the environmental implications of the smart grid. In addition the increased energy efficiency, demand response and load management will result in a stable development of clean power market.

7.7.1   Climate Change

Climate change is the term commonly linked to the issue of global warming and cooling resulting from the increased emissions of greenhouse gases (GHG). The term *climate change* refers to any distinct change in measures of climate lasting for a long period of time, that is, major changes in temperature, rainfall, snow, or wind patterns lasting for decades or longer. Climate change may result from:

- Natural factors, such as changes in the Sun's energy or slow changes in the Earth's orbit around the Sun.
- Natural processes within the climate system, for example, changes in ocean circulation
- Human activities that change the atmosphere's composition (for example, burning fossil fuels) and the land surface (for example, cutting down forests, planting trees, and expanding cities and suburbs)

Global warming is an average increase in temperatures near the Earth's surface and in the lowest layer of the atmosphere. Increases in temperatures in the Earth's atmosphere can contribute to changes in global climate patterns. Global warming can be considered part of climate change along with changes in precipitation, sea level, and so on.

Global change is a broad term that refers to changes in the global environment, including climate change, ozone depletion, and land use change.

7.7.2   Implications of Climate Change

The key implications of climate change include:

- Energy: Increased temperatures will cause an increase in energy bills as consumers use more air conditioning.
- Health: Extremes of temperature such as excessive and long-term exposure to heat will contribute to disease.
- Agriculture and Wildlife: Irregular weather variability implies lack of proper water supply, and increased temperature may result in worsening crop production and, ultimately, rising food costs.
- Water Resources: Temperature and weather irregularity increase the possibility of flooding and droughts and impact the quality and availability of global fresh water supply. As the water supply is affected, farmers will need to irrigate crops.

The DOE's strategic plan for a U.S. Climate Change technology program involves:

- The facilitation of DG, particularly RER technology
- The development of energy storage technology to conserve generated energy
- The advancement of DR to decrease demand

Consumer education and participation should increase sensitivity to the issues. Discussion of the potential contribution of RER on climate change will consider the projected increase, from the current 13% to 30%, as well as the following improvements to the grid:

   i. Improvement from 5% to 15% in DR systems

  ii. Improvement from 1% to 10% in consumer generation

 iii. Improvement from 47% to 90% of asset utilization

 iv. Improvement in transmission asset utilization from 50% to 80%

  v. Improvement in distribution asset utilization from 30% to 80%

7.8 STORAGE TECHNOLOGIES

Energy storage is important for utility load leveling, electrical vehicles, solar energy systems, uninterrupted power supply, and energy systems in remote locations. Energy storage has always been closely associated with solar installations, including solar heating and PV. Storage options are particularly essential when variable sources are used in islanding and standalone power systems. Figure 7.5 presents a sample topology for a microgrid system.

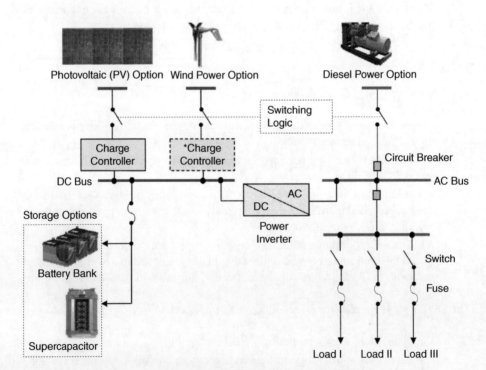

Figure 7.5. Microgrid topology with storage technologies.

There are two principal reasons why energy storage will grow in importance with the increased development of RER:

1. Many important RER are intermittent, generating when the weather dictates, rather than when energy demand dictates.
2. Many transportation systems require energy to be carried with the vehicle.

Storage options can be evaluated based on the characteristic of the application, for example, whether the application requires portable or fixed storage methods, the duration when storage will be operational, and the maximum power needed for the application. The selection of the proper storage technology is based on the following parameters:

- **Unit Size**: Scale of technology. Storage technologies have an associated range for application, for example, large units support grid-connected RER technologies.
- **Storage Capacity**: Total store of available energy after charging.
- **Available Capacity**: Average value of power output based on the state of charge/depth of discharge.
- **Self-discharge Time**: Time required for a fully charged, non-interconnected storage device to reach a certain depth of discharge (DOD), this is contingent on the operational condition of the system.
- **Efficiency**: Ratio of energy output from the device to the energy input issue of conversion technology and design of RER and storage and conversion needed.
- **Durability or Life-cycle**: Number of consecutive charge-discharge cycles a storage installation can undergo while maintaining the installations and other specifications within limited ranges. Life-cycle specifications are made against a chosen DOD depending on the applications of the storage device.
- **Autonomy:** Ratio between energy capacity and maximum discharge power; indicates the maximum amount of time the system can continuously release energy.
- **Mass and Volume Densities:** Amount of energy accumulated per unit mass or volume of the storage unit.
- **Cost:** Cost of installation, operation, and maintenance of storage technology; cost should be analyzed throughout system lifespan.
- **Feasibility:** Degree of adaptability to the storage applications.
- **Reliability:** Guarantee of service.

Additional information characteristics include monitoring and control equipment, operational constraints, environmental impacts, ease of maintenance, simplicity of design, operational flexibility, and response time for energy release. Table 7.2 compares the options.

TABLE 7.2. Comparison of Storage Technology Options

Storage Technology	Characteristics/Particulars	Advantages	Disadvantages
Flow Batteries	Similar to lead-acid batteries, but the electrolyte is stored in a external container and it circulates through the battery cell stack	• Unlimited electrical storage capacity, the only limitation is the size of the electrolyte storage reservoir	• Limited number of cycles of usage, after three (3) to five (5) years the system has to be changed
Advanced Batteries	Advanced batteries include lithium-ion, polymerion, nickel metal hybrid and sodium sulfur type	• Use less space than lead acid batteries	• Too expensive for large scale applications
Super capacitors	Electronic device with the capacity to provide high power and energy which have the characteristics of capacitors and electrochemical batteries except there is no chemical reaction.	• Virtually unlimited cycle life • Low impedance • Rapid charging • Simple charge methods	• Linear discharge voltage prevents use of the full energy spectrum • Low energy density • Cells have low voltages • High self-discharge

TABLE 7.2. (Continued)

Storage Technology	Characteristics/Particulars	Advantages	Disadvantages
Super Conducting Magnetic Energy Storage	Energy stored in the magnetic field created by the flow of direct current in a coil of superconducting material that has been cryogenically cooled.	• Power is available almost instantaneously • High power output for a brief period • No loss of power • No moving parts	• Energy content is small and short-lived • Cryogenics, cold temperature technology, can be challenging
Pumped Hydro	The process of water being pumped from a lower reservoir uphill then allowing it to flow downhill to through turbines to produce electricity	• Readily available and widely used in high power applications • Lower cost of power, frequency regulation on the grid, and reserve capability	Spends years in regulatory and environmental review • Can only be implemented in areas with hills
Compressed Air	Compressed Air Energy Storage (CAES) utilities use electricity generated during off-peak hours (i.e., storage hours) to compress air and store it in airtight underground caverns. When the air is released from storage, it expands through a combustion turbine to create electricity.	• Conserves some natural gas by using low-cost, heated compressed air to power turbines and create off-peak electricity	• Low efficiency due to the extra reheating energy needed to turn on the turbines • For every kilowatt-hour of energy going in, only .5 kilowatt-hour of energy can be taken out

(Continued)

TABLE 7.2. (*Continued*)

Storage Technology	Characteristics/Particulars	Advantages	Disadvantages
Flywheels	A cylinder that spins at a very high speed, storing kinetic energy. Utility — Rectifier — Bus — Converter — Load Motor/Generator — Bi-directional Inverter — Flywheel Vacuum Housing — Magnetic Bearings Hub — Composite Rim Courtesy of Beacon Power — 6 kWh unit	• Charge and discharge rapidly • Affected little by temperature fluctuations • Take up relatively little space • Long life span • Tolerant of abuse • Lower maintenance requirements than batteries	Power loss faster than for batteries

7.9 TAX CREDITS

Tax credits are examples of incentives to promote customer buy-in and participation in the utilization of RER and practice of energy efficiency. The DOE offers a number of tax credits to consumers [9]. These include:

- Home Energy Efficiency Improvement Tax Credit: Consumers who purchase and install specific products, such as energy-efficient windows, insulation, doors, roofs, and heating and cooling equipment in existing homes can receive a tax credit for 30% of the cost, up to $1,500.
- Residential Renewable Energy Tax Credit: Consumers who install solar energy systems (including solar water heating and solar electric systems), small wind systems, geothermal heat pumps, and residential fuel cell and microturbine systems can receive a 30% tax credit.
- Automobile Tax Credit for Hybrid Gas-Electric and Alternative Fuel Vehicles: Individuals or businesses who buy or lease a new hybrid gas-electric car or truck are eligible. Hybrid vehicles that use less gasoline than the average vehicle of similar weight and that meet an emissions standard qualify for the credit. Alternative-fuel vehicles, diesel vehicles with advanced lean-burn technologies, and fuel-cell vehicles are also eligible for tax credits.
- Automobile Tax Credit for Plug-In Electric Vehicles: The minimum amount of the credit is $2,500 and the credit tops out at $7,500, depending on the battery capacity.
- Automobile Tax Credit for Plug-In Hybrid Conversion Kits: The credit is equal to 10% of the cost of converting a vehicle to a qualified plug-in electric drive motor vehicle. The maximum amount of the credit is $4,000.

- Automobile Tax Credit for Low Speed & 2- or 3-Wheeled Vehicles: The Federal Recovery Act creates a special tax credit for certain low-speed electric vehicles and 2- or 3-wheeled vehicles. The amount of the credit is 10% of the cost of the vehicle, up to a maximum credit of $2,500.

7.10 SUMMARY

Chapter 7 has described progress in the utilization of cleaner, more environmentally responsible technologies for the electric system. Managing variability and interoperability were considered for the smart grid. Besides the increased utilization of RER, the chapter explained the need for conversion and storage technologies which must be further studied to determine the parameters necessary for optimal selection and implementation. New technologies covered included PHEV that advance transportation technology while providing additional storage for the smart grid.

REFERENCES

[1] J. Twidell and A.D. Weir. *Renewable Energy Resources*. USA: Taylor & Francis, 2006.

[2] R. Wiser and G. Barbose. "Renewables Portfolio Standards in the United States: A Status Report with Data Through 2007," *Lawrence Berkeley National Laboratory*, 2008.

[3] Thermal Energy System Specialists (TESS). The Transient Energy System Simulation Tool (TRNSYS).

[4] National Renewable Energy Laboratory 2010, Renewable Resource Data Center: PVWatts.

[5] J.I. Rosell and M. Ibanez. "Modeling Power Output in Photovoltaic Modules for Outdoor Operating Conditions," *Science Direct: Energy Conversion and Management* 2005, 472, 424–430.

[6] M. Kaltschmitt, W. Streicher, and A. Wiese, Eds., *Renewable Energy Technology, Economics and Environment*. New York: Springer, 2007.

[7] G. Boyle, "Renewable Electricity and the Grid: The Challenge of Variability," Earthscan, USA, 2007.

[8] C. Roe, J. Meisel, A.P. Meliopoulos, F. Evangelos, and T. Overbye. "Power System Level Impacts of PHEVs," *42nd Hawaii International Conference on System Sciences*, 2009, 1–10.

[9] W. Kempton and J. Tomic. "Vehicle to Grid Power Fundamentals: Calculating Capacity and Net Revenue," *Journal of Power Sources* 2005, 1–12.

SUGGESTED READING

National Renewable Energy Laboratory 2010, *HOMER: The Optimization Model for Distributed Power*.

8

INTEROPERABILITY, STANDARDS, AND CYBER SECURITY

8.1 INTRODUCTION

Deployment of the smart grid's components and interoperability requires a substantial overhaul of today's standards and protocols. In addition, improving the physical and cyber security of the network, which is notoriously vulnerable, is a top priority for the new architectural framework. Today's power distribution and monitoring are still in the initial stages of becoming a smart grid, with some substation network intelligence connected by microwave, power line, and/or fiber optic point to points. Although these core network backbones are very basic, they were never meant to securely connect two-way digital communication devices for every home, building, and appliance throughout a utility's service territory. In fact, adding millions of these connections to a distribution system is no easy task, and power companies are in the precarious position of having to prepare for the future.

If smart grids can realize their full potential, consumers, utilities, nations, and even the environment will benefit. Unfortunately, as with nearly any new technology, much of the focus has been on getting smart grid pilots and meters up and running, often with little consideration for cyber security. Worse, some experts appear to believe that the new grid's IT networks and industrial control systems can be secured by adapting exist-

Smart Grid: Fundamentals of Design and Analysis, First Edition. James Momoh.
© 2012 Institute of Electrical and Electronics Engineers. Published 2012 by John Wiley & Sons, Inc.

ing countermeasures. Therefore, the following discussion outlines the challenges for planners and designers and the role of policy-makers in attaining reliability and secure operations.

8.2 INTEROPERABILITY

Interoperability is "the ability of two or more systems or components to exchange information and to use the information that has been exchanged." The issues include the interoperation of system components supposedly conforming to a particular standard as well as the interoperation of components across standards. Merely having complete compliance to applicable smart grid standards is not enough to ensure interoperability. For example, the International Electrotechnical Commission (IEC) identifies the various objects and attributes as mandatory, optional, or conditional. Communications, management, security, and application execution messages must all be well understood by the interoperating equipment. A careful approach will include:

- Reviewing the activities of governing bodies: The outcome will determine the activities to be undertaken by smart grid users
- Reviewing components before deployment: Ensure compatibility with functional requirements
- Developing internal project standards: Address continuing issues and the governing body efforts.

8.2.1 State-of-the-Art-Interoperability

Interoperability is a key assumption in the development of smart metering to facilitate the competitive energy retail market [1]. There are two elements to interoperability: technical and commercial. Technical interoperability is largely about defining the functionality for gas and electricity metering interfaces providing smart metering (format and data content) service requirements. The definition of technical interoperability will depend on the market model the use of smart metering. WAN and LAN communications interfaces will be explored. Consumer engagement essential to delivering consumer benefits will be identified in a cost benefit analysis. Use of case studies will have to be forward looking to ensure that opportunities such as DSM and smart grids are not precluded by any solution.

8.2.2 Benefits and Challenges of Interoperability

Interoperability allows a network to seamlessly and autonomously integrate all components of electric power supply, particularly monitoring and measurement equipment, distribution and substation equipment, and management and communication equipment. The minimization of human intervention in this process is an important benefit of this functionality.

The challenges include the need for technical enhancement of the network, adoption and adaptation of existing technologies, and development and implementation of comprehensive standards. Procedures to address vandalism, hacking, and malicious attacks will require the development of security protocols for authentication and validation before access is granted.

8.2.3   Model for Interoperability in the Smart Grid Environment

Several key features in the interoperable efficiency of a network have to be considered in relation to each other and also independently [2]. The following illustrates a conceptual model of the smart grid developed by the GridWise Architecture Council (GWAC). An eight-layer stack, termed the GWAC stack, provides a context for determining smart grid interoperability requirements and defining exchanges of information [1]. The layers represent the chronological processes that enable various interactions and transactions within the smart grid. Each layer depends upon the layer beneath it and so each layer must function properly for the entire stack to be effective [2]. As more complex functions are required by the network, more layers will be required to achieve interoperability. According to GWAC, each category/driver subdivided by layers has a special purpose, as follows:

- Technical: Emphasizes the syntax or format of the information, focusing on how the information is represented on the communication medium
- Informational: Emphasizes the semantic aspects of interoperation, paying attention to what information is exchanged and its meaning
- Operational: Emphasizes the pragmatic (business and policy) aspects of interoperation, especially those pertaining to the management of electricity. [3]

8.2.4   Smart Grid Network Interoperability

Machines require specific data and instructions to complete tasks. The challenge is to design language and protocols to ensure effective communication between machines that are governed by the same protocol and even more so for those using different protocols. The main issue is to facilitate efficient, quick, and speedy transfer of data among and across devices. Interoperability is not limited to a physical aspect of the network, since design engineers must also consider that as two devices try to exchange data, the messages must now "speak the language" of network navigation and be properly "addressed" to reach the destination device.

This issue creates the need for networking standards so that machines that need to be connected can communicate without interruption or disruption. Unfortunately, this is not a simple task to undertake. For example, some machines may use a particular language protocol that requires another machine attempting to communicate to complete specific system requirements, which may be outside its scope, before transmission can be completed. In this case, the data transfer/communication may be prohibited or incomplete, making the network inefficient.

8.2.5 Interoperability and Control of the Power Grid

As discussed in the previous section, network control is a major issue for utilities. To upgrade to a smart system, the network should be outfitted with equipment that can detect problems, report back to the utility, receive control or restorative commands, and execute them. Full control will require that all machines communicate, interpret, and perform tasks that most machines today cannot do.

The use of SCADA and EMS have become ineffective, and leaves control centers with the need to communicate with other control centers as well as regulatory agencies, energy markets, independent power producers, large customers and suppliers, to keep up with the evolving market environment [6]. Control centers must be able to connect to machines that have smart technology to facilitate effective performance with little or no interruptions. Ultimately, the user/customer should have some degree of autonomy over consumption with a faster, more effective response to supply disruptions.

8.3 STANDARDS

Standards are the specifications that establish the fitness of a product for a particular use or that define the function and performance of a device or system [1]. Many standards bodies, including the National Institute of Standards and Technology (NIST), International Electrotechnical Commission (IEC), Institute of Electrical and Electronic Engineers (IEEE), Internet Engineering Task Force (IETF), American National Standards Institute (ANSI), North American Reliability Corporation (NERC), and the World Wide Web Consortium (W3C) are addressing interoperability issues for a broad range of industries, including the power industry. Table 8.1 provides a summary of the standards under development by key standards bodies.

The urgent need for the development of standards has led NIST to develop a plan to accelerate the identification and establishment of standards "while establishing a robust framework for the longer term evolution of the standards and establishment of testing and certification procedures." Based on the first phase of this work, NIST published the NIST Framework and Roadmap for Smart Grid Interoperability Standards Release 1.0 in September 2009 [8]. In this publication, nearly 80 existing standards are identified.

8.3.1 Approach to Smart Grid Interoperability Standards

The roadmap for interoperability by NIST includes the following applications:

- Demand Response and Consumer Energy Efficiency
- Wide Area Situational Awareness
- Electric Storage
- Electric Transportation
- Advanced Metering Management
- Distribution Grid Management

TABLE 8.1. Summary of Relevant Standards for Smart Grid Developed by Key Standards Bodies [4]

Standard Body	Description of Roles		Key Standards applicable to the Smart Grid Environment
The International Electrotechnical Commission (IEC)	Leading global organization which publishes standards for electrical electronic and related technologies for the electric power industry.	IEC 61850	• Substation automation, distributed generation (photovoltaics, wind power, fuel cells, etc.), SCADA communications, and distribution automation. Work is commencing on Plug—in Hybrid Electric Vehicles (PHEV).
	Applicable standards have been developed in the area of communication for the power industry.	IEC 61968	• Distribution management and AMI back office interfaces
		IEC 61850	• Substation automation, distributed generation (photovoltaics, wind power, fuel cells, etc.), SCADA communications, and distribution automation. Work is commencing on Plug—in Hybrid Electric Vehicles (PHEV)
		IEC 61968	• distribution management and AMI back office interfaces
		IEC TC 13 and 57	• Metering and communications for metering, specifically for AMI.
Institute of Electrical and Electronic Engineers (IEEE)	Standards in all areas of electrical, electronic and related technologies.	IEEE 802.3	• Ethernet
	Standards developed in the area of communications and interoperability.	IEEE 802.11	• WiFi
		IEEE 802.15.1	• Bluetooth
		IEEE 802.15.4	• Zigbee
		IEEE 802.16	• WiMax
Internet Engineering Task Force (IETF)	Responsible for Internet standards, dissemination of request for comment (RFC) documents for finalization of standards	RFC 791	• Internet Protocol (IP)
		RFC 793	• Transport Control Protocol (TCP)
		RFC 1945	• HyperText Transfer Protocol (HTTP)
		RFC 2571	• Simple Network Management Protocol (SNMP)
		RFC 3820	• Internet X.509 Public Key Infrastructure (PKI) for security

Organization	Description	Standards	Relevance
American National Standards Institute (ANSI)	Developed relevant standards for interoperability of AMI systems	ANSI C12.19 ANSI C12.22	• Metering "tables" internal to the meter • Communications for metering tables
National Institute of Standards and Technology (NIST)	Publications which provide guidelines toward secured interoperability.	NIST SP-800.53 NIST SP-800.82	• Recommended Security Controls for Federal Information Systems. • Guide to Industrial Control Systems (ICS) Security.
North American Electric Reliability Corporation (NERC)	Security standards for the bulk power system which may be extended to the distribution and AMI systems.	NERC CIP 002-009	• Bulk Power Standards with regards to Critical Cyber Asset Identification, Security Management Controls, Personnel and Training, Electronic Security Perimeter(s), Physical Security of Critical Cyber Assets, Systems Security Management, Incident Reporting and Response Planning, and Recovery Plans for Critical Cyber Assets
World Wide Web Consortium (W3C)	Interoperable technologies (specifications, guidelines, software, and tools) for the world wide web	HTML XML SOAP	• Web page design • Structuring documents and other object models • Web services for application-to-application communications for transmitting data

- Cyber Security
- Network Communications

These standards are applicable to transmission and/or distribution systems and pertain to electrical power system components. Table 8.2 presents such standards.

Additionally, there are several standards identified for different levels of communication, demand response, and measurement devices. As an example, in the area of power quality, the IEEE and other governing bodies have their own sets of standards.

8.4 SMART GRID CYBER SECURITY

Cyber security is a concept that has become increasingly prevalent with the development of the smart grid technology with the increased use of digital information and controls technology to improve reliability, security, efficiency of the electric grid and the deployment of smart technologies (real-time, automated, interactive technologies that optimize the physical operation of appliances and consumer devices) for metering, communications concerning grid operations and status, and distribution automation. The interaction of the power, communication, and information networks are critical to facilitating resiliency and sustainability of the infrastructures which further enhance the provision of adequate power and support economic and social growth of the nation. Technologies and protocols are developed for the maintenance of system, network, data, and SCADA security while conducting vulnerability assessment, incident recognition, recording, reporting, and recovery. Protection of network data as well as web-based or stored data is conducted.

8.4.1 Cyber Security State of the Art

Cyber security is a critical priority of smart grid development. However, the cyber security requirements for the smart grid are in a considerable state of flux. Cyber security includes measures to ensure the confidentiality, integrity, and availability of the electronic information communication systems necessary for the management and protection of the smart grid's energy, information technology, and telecommunications infrastructure.

This infrastructure includes information and communications systems and services and the information contained in these systems and services. Information and communications systems and services are comprised of the hardware and software that process, store, and communicate information, or any combination of all of these elements. Processing includes paper, magnetic, electronic, and all other media types.

Cyber security is defined as security from threats conveyed by computer or computer terminals and the protection of other physical assets from modification or damage from accidental or malicious misuse of computer-based control facilities [7]. Smart grid security protocols contain elements of deterrence, prevention, detection, response, and mitigation; a mature smart grid will be capable of thwarting multiple, coordinated

TABLE 8.2. Standards for the Various Electric Grid Levels

Level	Standard	Description	Application
Transmission /Distribution Level	IEEE Standards for Synchrophasors for Power Systems (IEEE C37.118-2005)	This standard defines synchronized phasor measurements used in power system applications. It provides a method to quantify the measurement, tests to be sure the measurement conforms to the definition, and error limits for the test. It also defines a data communication protocol including message formats for communicating this data in a real-time system.	Phasor measurement units communication
Transmission /Distribution Level	IEEE Standard for Interconnecting Distributed Resources with the Electric Power System (IEEE 1547-2003)	Itemizes criteria and requirements for the interconnection of distributed generation resources into the power grid.	Physical and electrical interconnections between utilized and distributed generation.
Transmission /Distribution Level	Common Information Model (CIM) for Power Systems (IEC 61968/61970)	Describes the components of a power system and power system software data exchange such as asset tracking, work scheduling and customer billing at an electrical level and the relationships between each component.	Application level energy management system interfaces.
Distribution	Communication networks and systems in substations (IEC 61850, Ed. 1 - 2009) IEC 61850	Standard for the design of electrical substations which addresses issues of interoperability, integration, intuitive device and data modeling and naming, fast and convenient communication. It includes abstract definitions of services, data and common data class, independent of underlying protocols.	Telecontrol /Telemetering Substation automation
Distribution /End User	Advanced Metering Infrastructure (AMI) System Security Requirements—AMI-SEC (June 2009)	Provides the utility industry and vendors with a set of security requirements for Advanced Metering Infrastructure (AMI) to be used in the procurement process, and represent a superset of requirements gathered from current cross-industry accepted security standards and best practice guidance documents.	Advanced metering infrastructure and SG end to end security
Distribution /End User	American National Standard For Utility Industry End Device Data Tables (ANSI C12.19-2008)	Defines a table structure for utility application data to be passed between an end device and a computer. Does not define device design criteria nor specify the language or protocol used to transport that data. The purpose of the tables is to define structures for transporting data to and from end devices.	Revenue metering information model

attacks over a span of time. Enhanced security will reduce the impact of abnormal events on grid stability and integrity, ensuring the safety of society and the economy.

There is a cyber security coordination task force directed at the design of grid security at the architectural level. Tasks include identifying used case studies with cyber security considerations, performing risk assessment such as vulnerabilities, threats and impacts as well as developing security architecture [2].

The underlying concept is that security should be built-in, not added-on. This strategy includes:

- Review of the system functionality and data flows with particular attention to their similarities and differences with identified dsmart grid use cases (as documented in the NIST Roadmap).
- Identification of relevant threats and the consequences/impacts if the confidentiality, integrity, availability, or accountability of the system data flows are compromised.

Security requires many different solutions and is not relegated to encryption and password protection. Facets of the cyber security include:

- Security assessment and hardening of the existing systems
- Vulnerability assessment
- Disaster recovery
- Intrusion detection incident response
- Event logging, aggregation, and correlation

Another critical understanding of the issue of security for the purpose of development is the realization of the inevitability of the occurrence of breaches. This leads to the development of contingency and recovery plans.

Table 8.3 itemizes the metamorphosis of the threats faced by the power system in the legacy power system grid and the new grid.

Critical objectives for the development of cyber security for the smart grid environment are ensuring the confidentiality, integrity, and availability of device and system data and communications channels, and securing logging, monitoring, alarming, and notification. Data protection will require confidentiality of communicated and stored data for the power system facilitated by authentication methods and the use of cryptography which includes encrypted authentication, to make it difficult for hackers. A combination of detective, corrective, and preventive controls will address cyber security risks. As the risks are determined, the best approach will balance efficiency, cost, and effectiveness.

All known cyber threats are monitored with a form of detective control. These include technologies such as host-based intrusion detection systems (HIDS) to monitor unauthorized changes to servers and deployed systems, network-based intrusion detection systems (NIDS) to detect network-based attacks, and platform-specific controls like virus detection and malware detection. These detective controls provide vital data

TABLE 8.3. Threats Facing the Electric Power System

	Traditional Threats faced by Legacy System	Threats faced by the New System
Impact	Direct damage to physical utility	Indirect damage to physical assets through damage to software systems
Location of origination of threat	Local	Local or remote
Target	Individuals	Individuals, competitors, and organizations
Point of Attack	Single site	Multiple point simultaneously
Duration of Damage	Immediate damage causing obvious damage	Attack may be undetected or lie dormant and then be triggered later
Occurrence	Single episode	Continued damage associated with attack
Restoration	Restoration after attack	Attacker may have continued impact preventing restoration

to design and testing staff about the real attacks and threats that the system faces. Detective controls often provide forensic-quality evidence that can be used to reconstruct attacks and potentially identify and/or prosecute attackers. Additionally, data from detective controls also assist in the selection of corrective and preventive controls.

Corrective controls seek to restore normal operations in the event of a successful cyber attack. Such controls are both manual, for example, a standardized procedure for switching to a backup system, and automatic, for example, failover designs that automatically disable compromised systems and replace them with known good systems. Corrections often seek to isolate and preserve successful attacks so that forensic analysis can proceed and permanent corrections, for example, in design, construction, or deployment, can be established.

For example, the secure development approach seeks to address risk throughout the life-cycle and throughout the engineering process, but preventive controls are a logical, efficient, and effective complement to secure implementation. Some attacks, for example, denial-of-service, can only be mitigated with a combination of secure engineering and preventive controls. Other attacks can be partially mitigated with a protective control, for example, network rules or role-based access controls, until a more comprehensive change to the system can be developed, tested, and deployed. As required, GridPoint deploys preventive controls to satisfy regulations, adhere to best practices, or to mitigate cyber security risks, including network-level filtering and rules, operating system-level mechanisms, and hardware-level mechanisms.

8.4.2 Cyber Security Risks

Cyber security risks appear in each phase of the project life-cycle and include risks to managerial, operational, and technical processes. These risks may impact equipment and systems, network management and integration, communications, control and operations, and system availability. The primary components that may be vulnerable to security risks include IT applications, communications network, and endpoints (for example, meters, in-home displays, and thermostats). There are substantial risks to the integrity of data and control commands due to the exchange of information through publicly accessible equipment, for example, smart meters using over-the-air communications technologies, for example, wireless or radio frequency, which may be intercepted and altered if not appropriately secured.

A number of system constraints need to be taken into account when satisfying security requirements. The requirements described do not prescribe which solution, for example, the use of narrow- or wide-band communications technologies, is most appropriate in a given setting. Such a decision is typically based on making prudent trade-offs among a collection of competing concerns. The following trade-off must include the considering cyber security risk:

- Other business or non-functional requirements
- Performance (for example, response time)
- Usability (for example, complexity of interactions for users)
- Upgradability (for example, ease of component replacement)
- Adaptability (for example, ease of reconfiguration for use in other applications)
- Effectiveness (for example, information relevant and pertinent to the business process as well as being delivered in a timely, correct, consistent, and usable manner)
- Efficiency (for example, the provision of information through the most productive and economical use of resources)
- Confidentiality (for example, protection of sensitive information from unauthorized disclosure)
- Integrity (for example, accuracy, completeness, and validity of information in accordance with business values and expectations)
- Availability (for example, information being available when required by the business process)
- Compliance (for example, complying with the laws, regulations, and contractual arrangements)
- Reliability (for example, the provision of appropriate information for management to operate the entity and exercise its fiduciary and governance responsibilities)

It is important to consider system constraints when developing applying security requirements, which include:

- Constraints
- Computational (for example, available computing power in remote devices)
- Networking (for example, bandwidth, throughput, or latency)
- Storage (for example, required capacity for firmware or audit logs)

- Power (for example, available power in remote devices)
- Personnel (for example, impact on time spent on average maintenance)
- Financial (for example, cost of bulk devices)
- Temporal (for example, rate case limitations)
- Technology
- Availability
- Maturity
- Integration/Interoperability (for example, legacy grid)
- Life-cycle
- Interconnectedness of infrastructure
- Applications (for example, automated user systems and manual procedures that process the information)
- Information (for example, data, input, processed and output by the information systems in whatever form is used by the business)
- Infrastructure (for example, technology and facilities, that is, hardware, operating systems, database management systems, networking, multimedia, and the environment that houses and supports them, that enable processing the applications)
- People (e.g., the personnel required to plan, organize, acquire, implement, deliver, support, monitor, and evaluate the information systems and services. They may be internal, outsourced or contracted as required) will consider: time, financial, technical, operational, cultural, ethical, environmental, legal, ease of use, regulatory requirements, scope/sphere of influence

8.4.3   Cyber Security Concerns Associated with AMI

AMI is the convergence of the power grid, the communications infrastructure, and the supporting information infrastructure [5]. This system of systems is constituted by a collection of software, hardware, operators, and information and has applications to billing, customer service and support, and electrical distribution. These applications each have associated cyber security concerns as summarized in Table 8.4.

The development of the security domain for AMI systems is addressed in Reference 5 and a security domain model was developed to bound the complexity of specifying the security required to implement a robust, secure AMI solution and to guide utilities in applying the security requirements to their AMI implementation. The services shown in Table 8.5 are descriptions of each of the six security domains. Each utility's AMI implementation will vary based on the specific technologies selected, the policies of the utility, and the deployment environment.

8.4.4   Mitigation Approach to Cyber Security Risks

This process of mitigation of errors or sources of insecurity includes the following:

- Identifying and classifying the information that needs to be protected
- Defining detailed security requirements

TABLE 8.4. Cyber Security Concerns Associated with AMI Systems

Application	Cyber-Security Concerns

Market Applications: Billing	• Confidentiality of:
	  ◦ Privacy of customer data, signals and location data
	• Integrity of:
	  ◦ Meter data
	  ◦ Signals for message and location and tamper indication
	• Availability of:
	  ◦ Meter data (for remote read), connect/disconnect service

Customer Applications	• Confidentiality of:
	  ◦ Access control for customer equipment via controls, price signals and messages
	  ◦ Privacy of customer data and payments
	• Integrity of:
	  ◦ Control messaging and message information containing prepayment data, usage data, rate information
	  ◦ Meter data for remote reading
	  ◦ Signals for message and location and tamper indication
	• Availability of:
	  ◦ Meter data (for remote read), connect/disconnect service, usage data, rate information
	  ◦ Customer payment data and usage balances customer devices

Distribution System Application	• Confidentiality of:
	  ◦ Access control of customer equipment including remote service switch and HAN devices
	• Integrity of:
	  ◦ Control messaging and message information
	  ◦ System Data
	• Availability of:
	  ◦ Customer devices
	  ◦ System data

TABLE 8.5. AMI Security Domain Descriptions

Security Domain	Description
Utility Edge Services	All field services applications including monitoring, measurement and control controlled by the utility
Premise Edge Services	All field services applications including monitoring, measurement and control controlled by the customer (the customer has the control to delegate to third party)
Communications Services	Applications that relay, route, and field aggregation, field communication aggregation, field communication management information
Management Services	Attended support services for automated and communication services (includes device management)
Automated Services	Unattended collection, transmission of data and performs the necessary translation, transformation, response, and data staging
Business Services	Core business applications (includes asset management)

- Reviewing the proposed security architecture that is designed to meet the requirements
- Procuring a system that is designed to meet the specified security requirements and includes the capability to be upgraded to meet evolving security standards
- Testing the security controls during the test and installation phase
- Obtaining an independent assessment of the security posture before deployment
- Developing a remediation plan to mitigate the risks for identified vulnerabilities
- Installing a system with built-in management, operational, and security controls
- Monitoring and periodically assessing the effectiveness of security controls
- Migrating to appropriate security upgrades as security standards and products mature
- Monitoring of communication channels
- Monitoring spike in usage (meter reading) to detect possible failures or tampering with the devices
- Making sure devices synchronize with the network within a given time frame to detect tampering, potential problems, and device failures.
- Penetration testing will be performed using the latest hacking techniques, to attempt to break into the systems, identifying possible vulnerabilities, and remotely validating the authenticity of the software running in the meters.

8.5 CYBER SECURITY AND POSSIBLE OPERATION FOR IMPROVING METHODOLOGY FOR OTHER USERS

Every communication path that supports monitoring and control of the smart grid is a two-way communication path. Each path is a potential attack path for a knowledgeable attacker. There are many potential entry points physically unprotected. Wireless networks can be easily monitored by attackers and may be susceptible to man-in-the-middle (MitM) attacks. Security mechanisms in place are intended to prevent unauthorized use of these communication paths, but there are weaknesses in these mechanisms. The history of security in complex networks implies that more vulnerability is yet to be discovered. Thus, the key points include:

1. Using spot checks on systems to go beyond the current paper chase approach to validating CIP compliance
2. Acknowledging that attackers and malware will find ways around/through current outer-wall–based network defenses, instituting a less-perimeter, defense-oriented approach to security controls with guidance on use of DMZs between internal networks.

8.6 SUMMARY

This chapter has delved into the fundamental tools and techniques essential to the design of the smart grid. The tools and techniques were classified into: (1) computational techniques and (2) communication, measurement, and monitoring technology. Based on the performance measures, that is, controllability, interoperability, reliability, adaptability, sustainability, efficiency, stochasticity, and predictivity, the chapter identified the most suitable applications of the tools. Ongoing work in the critical area of standards development by NIST and IEEE was explained, including consideration of the available standards to be adopted and/or augmented for application. The issue of interoperability was presented as it pertains to present grid technologies and the introduction of newer technologies. Acknowledging the grid's increasing dependence on communication and information systems is necessary to any discussion of the challenges of developing and deploying adequate cyber security protections.

REFERENCES

[1] J.A. Momoh. *Electric Power System Application of Optimization*, New York: Marcel Dekker, 2001.

[2] J.L. Marinho and B. Stott. "Linear Programming for Power System Network Security Applications," *IEEE Transactions on Power Apparatus and Systems* 1979, vol. PAS-98, pp. 837–848.

[3] R.C. Eberhart and J. Kennedy. "A New Optimizer Using Particle Swarm Theory," *Proceedings of the Sixth International Symposium on Micromachine and Human Science*, 39–31, 1995.

[4] G. Riley and J. Giarratano. *Expert Systems: Principles and Programming*, Boston: PWS Publisher, 2003.

[5] A. Englebrecht. *Computational Intelligence: An Introduction.* John Wiley & Sons, Ltd., 2007.

[6] M. Dorigo and T. Stuzle. *Ant Colony Optimization.* Cambridge, MA: Massachusetts Institute of Technology, 2004.

[7] "Appendix B2: A Systems View of the Modern Grid-Sensing and Measurement," National Energy Technology Laboratory, 2007.

[8] Report to NIST on the Smart Grid Interoperability Standards Roadmap—Post Comment Period Version. http://www.nist.gov/smartgrid/upload/Report_to_NIST_August10_2.pdf.

SUGGESTED READINGS

Z. Alaywan and J. Allen. "California Electric Restructuring: A Broad Description of the Development of the California ISO," *IEEE Transactions on Power Systems*, 1998, 13, 1445–1452.

A.G. Barto, W.B. Powell, D.C. Wunsch, and J. Si. *Handbook of Learning and Approximate Dynamic Programming.* IEEE Press Series on Computational Intelligence, 2004.

T. Bottorff. "PG&E Smart Meter: Smart Meter Program," *NARUC Summer Meeting*, 2007.

B. Brown. "AMI System Security Requirements," UCAIUG: AMI-SEC-ASAP, 2008.

M. Dorigo and T. Stutzle. "The Ant Colony Optimization Metaheuristic: Algorithms, Applications and Advances." In F. Glover and G. Kochenberger, eds.: *Handboook of Metaheuristics*. Norwell, MA, Kluwer, 2002.

B. Milosevic and M. Begovic. "Voltage-Stability Protection and Control Using a Wide-Area Network of Phasor Measurements," *IEEE Transactions on Power Systems* 2003, 18, 121–127.

J. Momoh, *Electrical Power System Applications of Optimization*, Boca Raton, FL: CRC Press, 2008.

"NIST Framework and Roadmap for Smart Grid Interoperability Standards," Office of the National Coordinator for Smart Interoperability, Release 1.0, 2009.

A.G. Phadhke. "Synchronized Phasor Measurements in Power Systems," *IEEE Computer Applications in Power* 1993, 6, 10–15.

D. Shirmohammadi, B. Wollenberg, A. Vojdani, P. Sandrin, M. Pereira, F. Rahimi, T. Schneider, and B. Stott. "Transmission Dispatch and Congestion Management in the Emerging Energy Market Structures," *IEEE Transactions on Power Systems* 1998, 13, 1466–1474.

H. Singh and F.L. Alvarado. "Weighted Least Absolute Value State Estimation Using Interior Point Methods", *IEEE Transactions on Power Systems* 1994, 9.

P.K. Skula and K. Deb. "On Finding Multiple Pareto-optimal Solutions Using Classical and Evolutionary Generating Methods," *European Journal of Operational Research* 2007, 181, 1630–1652.

"Smart Grid Communication Architecture," Pike Research, 2011.

"Smart Grid Initiatives White Paper," OSI Revision 1.1, 2009.

T.M. Smith and V. Lakshmanan. "Utilizing Google Earth as a GIS Platform for Weather Applications," *22nd International Conference on Interactive Information Processing Systems for Meteorology, Oceanography, and Hydrology*, 2006.

C.W. Taylor. "The Future in On-Line Security Assessment and Wide-Area Stability Control," *IEEE Power Engineering Society* Winter, 2000, 1, 78–83.

W.H. Zhange and T. Gao. "A Min-Max Method with Adaptive Weightings for Uniformly Spaced Pareto Optimum Points," *C4omputers and Structures* 2006, 84, 1760–1769.

L. Zhao and A. Abur. "Multiarea State Estimation Using Synchronized Phasor Measurements," *IEEE Transactions on Power Systems* 2005, 20, 2005.

D. Zhengchun, N. Zhenyong, and F. Wanliang. "Block QR Decomposition Based Power System State Estimation Algorithm," *ScienceDirect* 2005.

9

RESEARCH, EDUCATION, AND TRAINING FOR THE SMART GRID

9.1 INTRODUCTION

The smart grid will require engineers and professionals with greater expertise and training than the skilled workforce of today. In addition to the technological aspects of development [1], engineers will need to study manufacturing, data management, asset optimization, and policy and protocol development. The smart grid will also depend upon expanding current research efforts in the areas of cyber security, controls, communication, computational intelligence techniques, and decision support tools [2].

Operation and management of electricity generation, transmission, and distribution are changing due to technological and power marketing developments. Technological changes are being driven by the introduction of emerging technologies such as power electronics, DG, RER, microgrids, digital protection coordination, supervisory control, and energy management. The market-driven power business environment has created a need for nonelectrical engineering topics such as operations research and economics.

9.2 RESEARCH AREAS FOR SMART GRID DEVELOPMENT

Work is under way in developing research support for the intelligent grid in FACTS technology coordination and placement for real-time application, PMUs for real-time

Smart Grid: Fundamentals of Design and Analysis, First Edition. James Momoh.
© 2012 Institute of Electrical and Electronics Engineers. Published 2012 by John Wiley & Sons, Inc.

TABLE 9.1. Smart Grid Deployment

Problem	Classical method	Improvement Currently for static model	Recommendation for smart grid deployment
Stability	Lyapunov's Method; Transient energy function; Bifurcation method; Eigenvalue evaluation;	Quasi-steady state modeling;	Index for fast and accurate evaluation; Real time computation; using leaning algorithm and ADP for control coordination
OPF	Interior point method; Trust region method;	SQP, rSQP	ADP and heuristic, hybrid methods to account for prediction and stochasticity.
Economic Dispatch/ Unit Commitment	minimum cost while meeting the constraints;	Variants of nonlinear interior point methods;	Knowledge based system; Computational Intelligent method; to account for uncertainties and randomness
Reliability	State enumeration; Effects analysis; Reliability indices	Probabilistic input data for reliability study	Computational Intelligence and its hybrid method together with dynamics in the data

voltage stability, and reliability monitoring and control. Other research includes advanced distribution automation capable of handling DSM functions with real-time pricing options and mechanisms.

The research activities can be classified by the tools required (as shown in Table 9.1):

- **Simulation and analysis tools:** Simulate energy markets and energy systems and validate the vision. These tools will combine operations and economics in a single model to analyze and monitor the system as changes are implemented to determine the impacts and ensure fairness.
- **Development of smart technologies from the government and industry:** Jump-start the transformation
- **Testbeds and demonstration projects:** Provide experiments of increasing scale to prove the worth of these technologies or reveal their faults; build momentum for change, reduce the perception of risk, and build acceptance of the concept of a transformed energy grid [3].
- **New regulatory, institutional, and market frameworks:** Support a climate of innovation as technologies develop and evolve; fulfill the need for examination, provided to the network through education and development of research and training.

9.3   RESEARCH ACTIVITIES IN THE SMART GRID

The critical objectives of technical research include:

1. Develop advanced techniques for measuring peak load reductions and energy-efficiency savings from smart metering, DR, DG, and electricity storage systems
2. Investigate means for DR, DG, and storage to provide ancillary services
3. Conduct research to advance the use of wide-area measurement and control networks, including data mining, visualization, advanced computing, and secure and dependable communications in a highly distributed environment
4. Test new reliability technologies including those concerning communications network capabilities in a control room environment [4, 5] against a representative set of local and wide area outage scenarios
5. Identify communications network capacity needed to implement advanced technologies
6. Investigate the feasibility of a transition to time of use (TOU) and real-time electricity pricing
7. Develop algorithms for use in electric transmission system software applications
8. Promote the use of underutilized electricity generation capacity in any substitution of electricity for liquid fuels in the transportation system of the United States
9. Develop interconnection protocols to enable electric utilities to access electricity stored in vehicles to help meet peak loads

9.4   MULTIDISCIPLINARY RESEARCH ACTIVITIES

In addition to technical research, interdisciplinary research areas, such as economics, finance, policy, and environmental science, will incorporate aspects of development and implementation. Systems engineering courses such as intelligent systems and adaptive controls, pricing for new and emerging power markets, financial engineering, socioeconomics, and studies on climate change and environmental implications are all relevant to this discipline. The following lists some of the activities which demonstrate the need for interdisciplinary research and education:

- Development of advanced techniques for measuring peak load reductions and energy-efficiency savings
- Investigate means for demand response, distributed generation, and storage to provide ancillary services
- Conduct research to identify and advance the use of wide-area measurement and control networks
- Research on network asset management and architectures
- Test new reliability technologies, including those concerning communications network capabilities, in a grid control room environment

- Investigate the feasibility of a transition to TOU and real-time electricity pricing
- Development of algorithms for use in electric transmission system software applications

9.5 SMART GRID EDUCATION

Creating a professional development curriculum is a vital component for the future power industry. New areas of study, delivery tools, and professional development career paths need to be introduced periodically. The development of educational schemes will require skills and technologies outside of power engineering. Development of infrastructure to facilitate the delivery of courses such as the development of a Web-based information data bank, interactive and distant power laboratory experiments, and encouraging international cooperation for knowledge and expertise exchange are necessary.

Smart grid fundamentals will include definitions, architecture, metrics of performance requirements, discussion of development of analytical and decision support tools, as well as RER. Grid design will be based on cross boundaries of knowledge in communication theory, optimization, control, social and environmental constraints, and dynamic optimization techniques.

Conventional Electrical Engineering courses are directed toward the development of engineers for jobs relating to the operation of the electric power distribution and telecommunications infrastructure, designing communication systems, and developing electronics or control systems. The power system curriculum allows for an introduction to theory or practical work in many of these areas with a major focus on the following fundamental modules of a current modular power engineering program [1]:

- Utilization and management of power and energy
- Application of new emerging technologies for power quality
- High voltage engineering; power electronics control
- Multidisciplinary power engineering related topics

Desired elements of the curriculum have been identified to facilitate the development of the smart grid include:

  (i) direct digital control
 (ii) roles of system operators
(iii) power systems dynamics and stability
 (iv) electric power quality and concomitant signal analysis
  (v) transmission and distribution hardware and the migration to middleware
 (vi) new concepts in power system protection
(vii) environmental and policy issues

(viii) reliability and risk assessment

(ix) economic analysis, energy markets, and planning

Course development can be established at the senior undergraduate level as well as for graduate students. Outside the university environment, modules can be geared toward the technical staff of power systems and policy-makers. Integration of technical components aimed at providing exposure to the smart grid design is needed. The new curriculum requires the following key features:

- Planning and operation under uncertainties
- Use of real-time measurements, techniques and tools such as PMUs and SE for analysis of stability, reliability, and efficiency.
- Renewable energy with vulnerability and penetration strategies and associated storage technologies
- Performance measures and issues which include sustainability, power quality, interoperability, and cyber security
- Development of new adaptive and stochastic optimization techniques that will facilitate resource allocations and scheduling such as unit commitment, restoration, and reconfiguration
- Marketing and pricing ancillary services and business cases for smart grid deployment

A sample syllabus for a smart grid fundamentals course will be divided into several modules [2].

9.5.1 Module 1: Introduction

- What is the smart grid?
- Working definitions and associated concepts
- Smart grid functions

This module will introduce the smart grid along with a history of the development of the smart grid including the contributions of key players and policy developments. A working definition and characteristics will also be included.

9.5.2 Module 2: Architecture

- Components and Architecture of Smart Grid Design

This module will review the proposed architecture. The fundamental components of smart grid designs will include: transmission automation, system coordination, situation assessment, system operations, distribution automation, renewable integration, energy efficiency, DG and storage, demand participation signals and options, smart appliances, PHEVs, and storage.

9.5.3   Module 3: Functions

- Review the functions of the smart grid
- Evaluation of components: generation, transmission, distribution, and end-user

This module will partition the smart grid with layers of intelligence and functions by technology, new tools, and innovations in energy supply and consumption for efficiency and survivability.

9.5.4   Module 4: Tools and Techniques

- Computational Techniques
  - Analytic methods (decision support tools, static and dynamic optimization techniques)
- Computational Intelligence Techniques
- Introduction to Technologies
  - Communication technology
  - Sensing, metering, and measurement technologies
  - RER

This module will introduce classical tools and techniques, new global optimization techniques, and cyber technology for anticipatory, predictability, and adaptability to sustainability and resiliency of the infrastructure and its interdependencies. Evaluation of the tools will be conducted considering the performance measures for their development, particularly controllability, interoperability, reliability, adaptability, sustainability, anticipatory behavior, and security.

9.5.5   Module 5: Pathways to Design

- Selection criteria for tools and techniques
- Advanced optimization and control techniques
- Automation at generation, transmission, distribution, and end-user levels

This module will apply advanced optimization techniques for the automation of various grid level functions.

9.5.6   Module 6: Renewable Energy Technologies

- Introduction
- Storage Technologies

- EVs and plug-in hybrids
- Environmental impacts and climate change
- Economics

This module will introduce renewable energy technologies and study the sources, storage and electronic technologies, as well as EVs and plug-in hybrids. Characteristics, advantages and disadvantages of these technologies will be discussed and the associated environmental and economic issues emphasized.

9.5.7   Module 7: Communication Technologies

- Introduction Network topologies
- WAMS
- AMI

This module will discuss communication technology for the development of open architecture that securely integrates smart sensors and control devices, control centers, protection systems, and end-users.

9.5.8   Module 8: Standards, Interoperability, and Cyber Security

This module will involve studies on the development and implementation of standards, interoperability, and cyber security for the smart grid environment.

9.5.9   Module 9: Case Studies and Testbeds

This module will present case studies of generation, transmission, and distribution. Implementation of various technologies at industry locations and the establishment and utilization of academic and research testbeds will be included. Instruction will extend beyond the lecture format, with project based learning (PBL), case studies, and experiments.

9.6   TRAINING AND PROFESSIONAL DEVELOPMENT

For several years, the power industry has faced the realization that the trained workforce is aging much like the system itself. Smart grid deployment is an opportunity to address these two issues at once. Programs in place to upgrade the skills of existing workers and to train future workers include the new Digital and Distributed Power Systems Training Program. This expertise entails that both operators and engineers have high coordination skills including multidisciplinary knowledge [5].Training and re-education must equip current and potential employees at all levels of grid development with knowledge of advanced cyber security tools and technologies for critical controls systems, which the Department of Homeland Security has identified as top priorities.

Industry experts and regulatory bodies are unanimous in stating that resilience and cyber security are critical areas of concern for the future grid.

9.7 SUMMARY

This chapter has highlighted research and education in a proactive approach for achieving a robust, scalable power system network development. Nine educational modules to promote smart grid awareness and enhancement were described. The importance of upgrading workforce skills and training the future workforce was emphasized.

REFERENCES

[1] G. Heydt et al. "Professional Resources to Implement the 'Smart Grid,'" *NAPS Power Symposium*, 2009.

[2] J. Momoh. "Fundamentals of Analysis and Computation for the Smart Grid," *IEEE PES General Meeting*, 2010.

[3] J. Rosell and M. Ibanez. "Modeling Power Output in Photovoltaic Modules for Outdoor Operating Conditions," *Science Direct: Energy Conversion and Management* **2005**, 47, 2424–2430.

[4] M. Kaltschmitt, W. Streicher, and A. Wiese, eds. *Renewable Energy Technology, Economics and Environment*. Springer, 2007.

[5] G. Boyle. *Renewable Electricity and the Grid: The Challenge of Variability*, Earthscan, 2007.

10

CASE STUDIES AND TESTBEDS FOR THE SMART GRID

10.1 INTRODUCTION

Much work has been undertaken in the areas of policy and standards, framework development, and the assessment of new, intelligent tools to achieve the objectives of the smart grid. With the financial support of utilities and government agencies, the work is primarily directed to the preparation of the smart grid environment, particularly in terms of retrofitting and increasing the efficiency of the grid. Installation of WAMs and smart metering upgrades are two additional areas to which utilities are committing time and resources. The development of tools, packages, demonstration environments, and testbeds is important for the research, education, and technological growth that will support successful deployment.

10.2 DEMONSTRATION PROJECTS

Demonstration projects related to the development of WAMS and voltage monitoring were completed by the DOE feasibility and design of WAMS [1] as early as 1996. DOE stimulus funding has been provided for numerous utilities to engage in technical

Smart Grid: Fundamentals of Design and Analysis, First Edition. James Momoh.
© 2012 Institute of Electrical and Electronics Engineers. Published 2012 by John Wiley & Sons, Inc.

manufacturers upgrade, and research support to academic institutions. Projects using smart meter and distribution technology and consumer/end-user technology integration and deployment are organized and used by some utilities. [3]

Smart grid stimulus funds awarded by DOE [2] for 2009 totaled $620 million. Two categories of demonstration projects were funded: (i) fully integrated regional smart grid demonstrations for communication technologies, sensing and control devices, smart meters and in-home monitoring systems, energy storage options, and renewable energy integration; and (ii) utility-scale energy storage projects.

Commonwealth Edison, Dominion Virginia Power, and Duke Energy were among the utilities receiving $10–200 million primarily for smart meters throughout the United States. Southern California Edison and several other companies received grants for synchrophasor installation, wind energy storage, DR, distribution automation, and regional smart grid demonstration projects. The GridComm project by San Diego Gas & Electric in conjunction with Cisco, IBM, and Arcadian Networks developed a wireless smart grid network.

10.3 ADVANCED METERING

Utilities such as Pepco are investing in the installation and integration of smart meters. The Pacific Northwest Smart Grid Demonstration Project aims to validate new smart grid technologies and business models and provide two-way communication between distributed generation, storage, demand assets, and the existing grid infrastructure while quantifying smart grid costs and benefits and advanced standards for interoperability and cyber security approaches.

10.4 MICROGRID WITH RENEWABLE ENERGY

The feasibility of a microgrid as a standalone system will be improved if it can generate enough power for the load it is designed for. Microgrid development is a customer system-oriented project meant to enable smart grid technologies and functions for buildings and facilities. The development of an integrated system encompasses the implementation of smart grid equipment, devices, and software, as well as installation and the improvement of advanced smart meter technologies into a community to impact the distribution system. In most implementations, microgrid development encompasses the retrofitting infrastructure and DER resource installation within the identified system. The intellectual research which will support this collaborative will involve an economic and benefit analysis, a platform for the analysis of network performance which constitutes the system level of the intelligent microgrid in addition to a platform for decision support.

Figure 10.1 depicts a sample testbed topology design for such a study, with two PV and wind technology. In the microgrid topology [2], storage technologies are necessary, particularly for islanded or stand-alone operation of the microgrid.

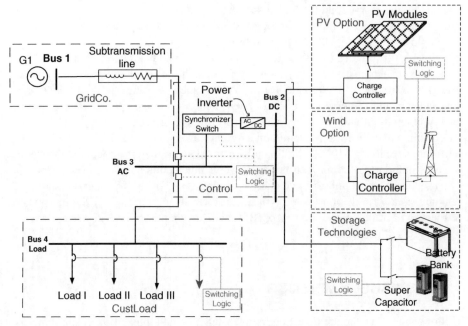

Figure 10.1. Sample microgrid testbed environment.

10.5 POWER SYSTEM UNIT COMMITMENT (UC) PROBLEM

The objective function of the UC problem can be formulated as the sum of the costs of all units over time, and presented mathematically as [4, 5]:

$$F = \sum_{t=1}^{T} \sum_{i=1}^{N} [u_i(t)F_i(t)) + S_i(t)]$$

The constraint models for the UC optimization problem are as follows:

- System energy balance

$$0.5 \sum_{i=1}^{N} [u_i(t)P_{g_i}(t) + u_i(t-1)P_{g_i}(t-1)] = P_D(t)$$

- Energy and power exchange

$$E_i(t) = 0.5[P_{g_i}(t) + P_{g_i}(t-1)]$$

- Spinning reserve requirements

$$\sum_{i=1}^{N} u_i(t)P_{gi}(t) \geq P_D(t) + P_R(t)$$

- Unit generation limits

$$P_{gi}^{\min} \leq P_{gi}(t) \leq P_{gi}^{\max}$$

With $t \in \{1,T\}$ and $i \in \{1,N\}$ in all cases where:

F : Total operation cost on the power system

$E_i(t)$: Energy output of the ith unit at hour t

$F_i(E_i(t))$: Fuel cost of the ith unit at hour t

$u_i(t)$: Ratio of generation output and capability

N : Total number of units in the system

T : Total time under which UC is performed

$P_{gi}(t)$: Power output of the ith unit at hour t

$P_{gi}^{\max}$: Maximum power output of the ith unit

$P_{gi}^{\min}$: Minimum power output of the ith unit

$S_i(t)$: Start-up cost of the ith unit at hour t

In the reserve constraints, the classifications for reserve include units on spinning reserve and units on cold reserve under the conditions of banked boiler or cold start.

Lagrangian relaxation is commonly used to solve UC problems. It is much more beneficial for utilities with a large number of units since the degree of suboptimality goes to zero as the number of units increases. It is easily modified to model characteristics of specific utilities and It is relatively simple to add unit constraints. The drawback of Lagrangian relaxation is its inherent suboptimality.

$$L(\lambda, \mu, \nu) = \sum_{t=1}^{T} \sum_{i=1}^{N} [C_i(P_{gi}(t)) + S_i(x_i(t))] + \lambda(t)\left(P_d(t) + P_R(t) - \sum P_{gi}\right) + \mu(t)(P_{gi}^{\max} - P_{gi})$$

where $\lambda(t)$, $\mu(t)$ are the multipliers associated with the requirement for time t.

Solution Approach using ADP Variant for the UC Problem: ADP is able to optimize the system over time under conditions of noise and uncertainty. If optimal operation samples are used to train the networks of the ADP, the neural network can learn to commit or adapt the generators, follow the operators' patterns, and change the operation according to the load changing. Figure 10.2 is a schematic diagram for implementations of heuristic dynamic programming (HDP). The input of the action network is the states of the generators and the action is how to adjust their output. Output J represents the cost-to-go function and the task is to minimize the J function.

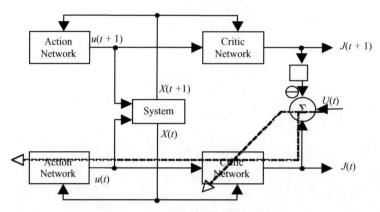

Figure 10.2. Scheme of implementation of HDP.

In this diagram, the input is the state variable of the network, and it is the cost of the generation vector presented as $X = [C(P_{gi})]$. The output is the control variables of units, and it is the adjustment of unit generation presented as: $u = [\Delta P_g]$. The utility function is local cost, so it is a cost function about unit generation within any time interval presented as $U = f(P,t)$. After transposing the power system variables using these guidelines, the scheme of implementation of HDP include the following computations.

The error of the critic network is:

$$e_C(t) = \gamma J(t) - J(t+1) - U(t)$$

and the updating weight using:

$$w_C(t+1) = w_C(t) + \Delta w_C(t)$$

and

$$\Delta w_C(t) = \eta e_C \left[-\frac{\partial e_C(t)}{\partial w_C(t)} \right]$$

where

$$\frac{\partial E_C}{\partial w_{Cij}^{(1)}} = \frac{\partial E_C}{\partial e_C} \cdot \frac{\partial e_C}{\partial y_{Ck}} \cdot \frac{\partial h_{Ck}}{\partial h_{Ck}'} \cdot \frac{\partial h_{Cj}'}{\partial w_{Cij}^{(1)}}$$

$$= \gamma e_C \cdot \left[\frac{1}{2}(1 - h_{Cj}^2) \right] \cdot w_{Cj}^{(2)} x_i$$

$$\frac{\partial E_C}{\partial w_{Cjk}^{(2)}} = \frac{\partial E_C}{\partial e_C} \cdot \frac{\partial e_C}{\partial y_{Ck}} \cdot \frac{\partial y_{Ck}}{\partial w_{Cjk}^{(2)}} = \gamma e_C y_{Ck}$$

I : Number of elements in R vector
J : Number of hidden layer nodes
K : Number of output layer nodes
M : Number of elements in u (action) vector
h'_C : Hidden layer input nodes
h_C : Hidden layer output nodes
y'_C : Output layer input nodes
y_C : Output layer output nodes
$w_C^{(1)}$: Weights between input and hidden layers
$w_C^{(2)}$: Weights between hidden and output layers
x : Input layer nodes

The error of the action network is computed as:

$$e_A(t) = J(t) - U(t)$$

and the updating weight is

$$w_A(t+1) = w_A(t) + \Delta w_A(t)$$

and

$$\Delta w_A(t) = \eta e_A \left[-\frac{\partial e_A(t)}{\partial w_A(t)} \right]$$

where

$$\frac{\partial E_A}{\partial w_{Ajk}^{(2)}} = \frac{\partial E_A}{\partial e_A} \cdot \frac{\partial e_A}{\partial J_k} \cdot \frac{\partial J_k}{\partial y_{Ak}} \cdot \frac{\partial y_{Ak}}{\partial y'_{Ak}} \cdot \frac{\partial y'_{Ak}}{\partial w_{Ajk}^{(2)}}$$

$$= \gamma e_A h_{Aj} \cdot \left[\frac{1}{2}(1 - h_{Aj}^2) \right] \cdot \left[\sum_{j=1}^{J} w_{Cj}^{(2)} \frac{1}{2}(1 - h_{Cj}^2) w_{Cij}^{(1)} \right]$$

$$\frac{\partial E_A}{\partial w_{Aij}^{(1)}} = \frac{\partial E_A}{\partial e_A} \cdot \frac{\partial e_A}{\partial J_k} \cdot \frac{\partial J_k}{\partial y_{Ak}} \cdot \frac{\partial y_{Ak}}{\partial y'_{Ak}} \cdot \frac{\partial y'_{Ak}}{\partial w_{Aij}^{(1)}}$$

$$= \gamma e_A w_{Ajk}^{(2)} x_i \cdot \left[\frac{1}{2}(1 - h_{Aj}^2) \right] \cdot \left[\frac{1}{2}(1 - y_{Ak}^2) \right] \cdot \left[\sum_{j=1}^{J} w_{Cj}^{(2)} \frac{1}{2}(1 - h_{Cj}^2) w_{Cij}^{(1)} \right]$$

The corresponding calculation steps are as follows:

Step 1: Use the sample data to pretrain the action network. The error is the difference between the output and the real value.

Step 2: Use the sample data to train the critic network with the pre-trained and unchanged action network. Use Equations (7) ~ (12) to update the weights. Then begin to apply the mature ADP network in the real work.

Step 3: Input the current state data $X(t)$ to the action network.

Step 4: Get the output $u(t)$ of the action network. Input $u(t)$ to the system function to get the state of next time $X(t + 1)$.

Step 5: Use the state of next time $X(t + 1)$ to get the action of next time $u(t + 1)$.

Step 6: Input the action and state of different time $u(t)$, $X(t)$ and $u(t + 1)$, $X(t + 1)$ to different critic network, respectively, and J functions for different time $J(t)$, $J(t + 1)$ are obtained.

Step 7: Back propagate and update the weights of the critic and action networks. Then time $t = t + 1$.

Results: Figure 10.3 from Reference 6 shows the structure of the neural network in HDP for a 3-generator system.

Figure 10.4 shows the load curve of a 3-generator, 6-node system. The closeness of the line graphs indicates that the ADP method generates correct results.

After training, the HDP can give the generation plan, which is very close to the optimal plan. The HDP [4] method can deal with the dynamic process of UC, and easily find a global optimal solution, which is difficult for classical optimization methods. Figure 10.5 shows the generation schedule of a 3-generator system.

In Figure 10.5, X1, X2, and X3 present the output of the three generators, respectively, and [X1], [X2], and [X3] present their expected (or optimal) output.

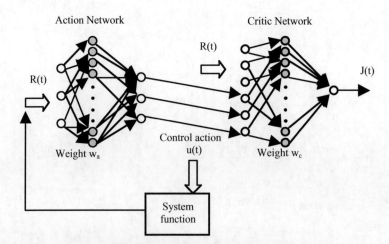

Figure 10.3. Structure of the neural network in HDP.

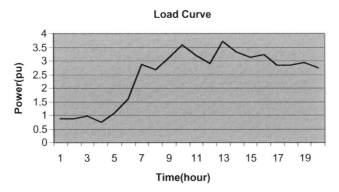

Figure 10.4. Load curve of 3-generator, 6-node system.

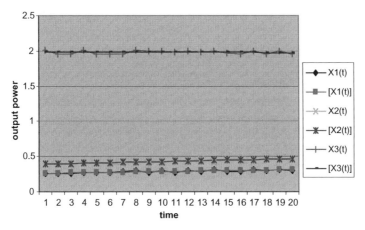

Figure 10.5. Generation schedule for the UC problem solved using ADP.

10.6 ADP FOR OPTIMAL NETWORK RECONFIGURATION IN DISTRIBUTION AUTOMATION

Distribution networks are generally configured radially for effective and noncompli-cated protection schemes. Under normal operating conditions, distribution feeders may be reconfigured to satisfy the objectives of minimum distribution line losses, optimum voltage profile, and overloading relief. The minimum distribution line loss is formulated as:

$$Minimize \sum |z_b i_b|$$

subject to:

$$[A]i = I$$

where:

Z_b : Impedance of the branch

I_b : complex current flow in branch b

i : m-vector of complex branch currents

A : $n \times m$ network incidence matrix, whose entries are

    = +1 if branch b starts from node p

    = −1 if branch b starts from node b

    = zero if branch b is not connected to node p

m : Total number of branches

n : Total number of network nodes

I : n-vector of complex nodal injection currents

The illustrative example problem is solved using the integer interior point method presented in Reference 7. Figure 10.8 shows the ADP method for a 5-bus system.

It involves the development of an ADP framework which involves (a) action network, (b) critic network, and (c) the plant model as shown in Figure 10.6 for network distribution reconfiguration.

The algorithm to solve this problem using ADP is presented in Figure 10.8.

To solve the optimal distribution reconfiguration problem using the ADP algorithm, we need to model the four parts of system structure shown in Figure 10.8: action vectors, state vectors, immediate rewards, and the plant. The system is tested with a 5-bus and a 32-bus system. We discuss the different parts of the ADP implementation structure as follows:

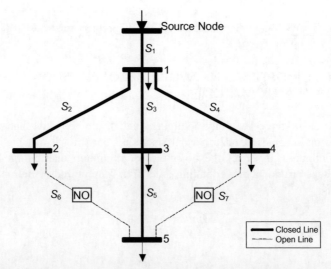

Figure 10.6. Small power system for network distribution reconfiguration problem.

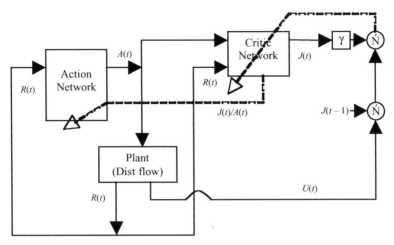

Figure 10.7. ADP structure network reconfiguration.

Rewards (Utility function): Optimal reconfiguration involves selecting the best set of branches to be opened, one from each loop, such that the resulting radial distribution system has the desired performance. Amongst the performance criteria considered for optimal network reconfiguration, the one selected is the minimization of real power losses. Application of the ADP to optimal reconfiguration of radial distribution systems is linked to the choice of an immediate reward U, such that the iterative value of J is minimized, while the minimization of total power losses is satisfied over the whole planning period. Thus, we compute the immediate reward as:

$$U = -Total\ Losses$$

Action vectors: If each control variable A_i is discretized in d_{u_i} levels (for example, branches to be opened one at each loop of radial distribution systems), the total number of action vectors affecting the load flow is:

$$A = \prod_{i=1}^{m} d_{u_i}$$

where m expresses the total number of control variables (for example, the total number of branches to be switched out).

The control variables comprise the sets of branches to be opened, one from each loop. From the network above, we can easily deduce from the simple system the entire set of action vectors that can maintain the radial structure of the network. The combinations are:

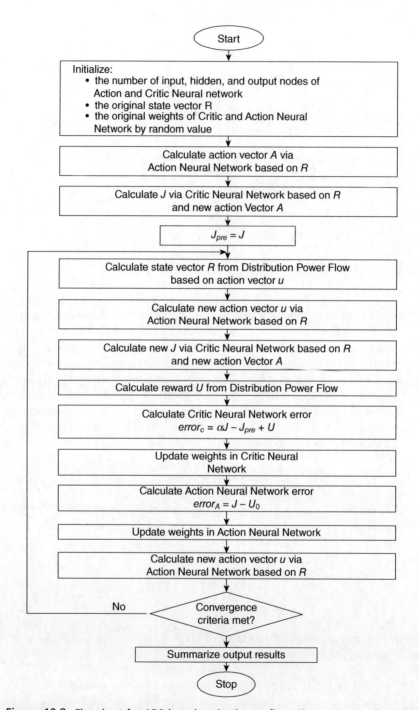

Figure 10.8. Flowchart for ADP-based optimal reconfiguration strategy using ADP.

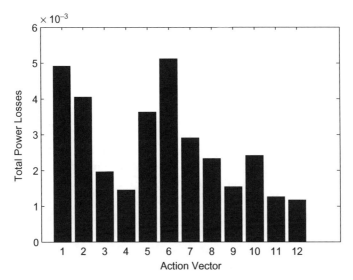

Figure 10.9. Action vector performance during system training.

A_1: {open switches 2, 3; close all other switches}

A_2: {open switches 6, 3; close all other switches}

A_3: {open switches 2, 5; close all other switches}

A_4: {open switches 6, 5; close all other switches}

A_5: {open switches 2, 4; close all other switches}

A_6: {open switches 3, 4; close all other switches}

A_7: {open switches 6, 4; close all other switches}

A_8: {open switches 5, 4; close all other switches}

A_9: {open switches 2, 7; close all other switches}

A_{10}: {open switches 3, 7; close all other switches}

A_{11}: {open switches 6, 7; close all other switches}

A_{12}: {open switches 5, 7; close all other switches}

Some Simulation Results: The purpose of the algorithm is to find the optimal switches status combination; for the 5-bus case, the optimal solution is Action Vector 15.

Figure 10.9 shows the minimization of the losses as action vectors for the optimal switching sequence. After the initialization, the action network generates the first action vector by random number. The action vector then inputs into the critic vector with state variables. The output of critic network J and immediate cost U obtains the new error for the action and critic networks. The weights in the two networks can be updated based on back propagation rules. After sufficient iterations, the system will output the result, in our case, the optimal action vector, which is the best switching status combination with the minimum losses.

We recommend extending this study to a large-scale aerospace power system while addressing the multiobjective challenges of restoration, reconfiguration, and remedial control.

10.7 CASE STUDY OF RER INTEGRATION

1. AEP, which has its own existing EHV 765 kilovolt (kV) network, consisting of 19,000 miles of transmission lines, would establish a combined additional capacity of perhaps 200–400 GW of bulk transmission, thus spurring significantly increased levels of wind energy in the overall energy portfolio. AEP estimated the cost in the range of $60 billion in 2007.
2. Cape Wind took one step closer to becoming the first offshore wind producer in the United States when Massachusetts regulators approved a power purchase agreement with National Grid for fifty percent of the output. The state's Supreme Judicial Court is expected to rule on the PPA by December 31, 2011.
3. Southern California Edison is moving toward smaller PV plants as the price of solar panels continues to drop. It recently signed contracts for 239.5 MW to be provided by 20 small solar farms.
4. A U.K.-based electricity company is developing an electric sports car that gets its power from wind turbines, to break the British land speed record of 139 mph in 2012.
5. With 45 wave and tidal prototypes slated for ocean testing this year and next, IHS Emerging Energy Research says the global ocean power market looks poised for takeoff.
6. The U.S. Navy is planning to roll out about 10 ships, planes and submarines powered by a blend of biofuels and nuclear power.

10.7.1 Description of Smart Grid Activity

Two important ways that smart grid technology creates value in the transmission network are

- Direct interconnection of large-scale resources
- Using remote sensing (and eventually, remote devices for the injection of dynamic reactive power) to increase network capacity by improving flow management

10.7.2 Approach for Smart Grid Application

Advanced metering infrastructure (AMI). AMI is a catch-all term for smart meters. The smart grid is often associated with electrical meters, or smart meters. This association helps to simplify a very complex process, but it sells the vision of the smart grid short.

The smart grid is no more about the meter than the economy is only about the cash register. Both are merely collection points: interfaces between buyer and seller where data can be gathered and analyses. As in many business and technology scenarios, you can't improve what you can't measure. AMI systems capture data, typically at the meter, to provide information to utilities and transparency to consumers. It's also usually the case that AMI systems piggyback on a variety of wireless systems. One simple illustration: the days of the electric company representative swinging by to read your meter could be phased out with AMI.

Demand response (DR). To date, consumers have used energy whenever they want to, and utilities have built the power plants and delivery infrastructure to support it, no matter what the cost or environmental impacts. To achieve economic and environmental goals, consumers need to become equal participants in the process, tuning their energy consumption to when clean resources are available and avoiding peak energy consumption times as much as they can. If some electricity-consuming devices can be deferred to nonpeak time, everyone wins. While you may need to turn on your lights when you arrive home, there's no reason that you can't run your dishwasher at 3 AM when rates and demand are lower.

That's the reasoning behind demand response programs and they have been very successful. Research bears out again and again that when consumers are asked to "do the right thing" when it comes to energy usage, they will do so.

Smart thermostats, for example, can prompt consumers to lower their air conditioner by a degree or two. To date, [5] some U.S. citizens are in some kind of demand response program. In-home displays and similar devices can lower energy use by up to 6%. Demand response programs yield immediate reductions on the average.

Critical peak pricing (CPP). An off-shoot of demand response, critical peak pricing simply means that utilities have the technology infrastructure in place to charge consumers more for energy during peak periods. It allows customers to decide whether or not to pay more on the specific critical days, rather than paying an average cost. It helps balance cost and risk between the consumer and the utility, as well as providing a further incentive for consumers to reduce energy consumption.

Time-of-Use Pricing (TOU). Time-of-use pricing is similar to critical peak pricing, except extrapolated across every hour for every day. Time-of-use pricing allows utility rates and charges to be assessed based on when the electricity was used. Not only the time of day, but also the season, as well as accounting for local weather patterns that might prompt a rate adjustment (at which point it becomes a demand response application).

10.8 TESTBEDS AND BENCHMARK SYSTEMS

The development of computational tools for power system applications requires extensive testing and validation for efficiency, speed, accuracy, reliability, and robustness.

We will require data and/or users to test the final product based on the uniqueness of the test system being studied in normal, alert, emergency, and restoration conditions, with approved standards for interoperability, cyber security, and managing uncertainty.

We hope to continue development of a MATLAB-based [4, 5] environment for generalizing the dynamic stochastic OPF, the variants of ADP, decision analysis tools, and other tools for solving different testbeds and benchmarks in civilian and military power networks. The results will be discussed a future book entitled, "Adaptive Essential Optimization Power Techniques (AOPT)," which will provide a development platform for smart grids and other complex system networks.

10.9   CHALLENGES OF SMART TRANSMISSION

Two key issues in developing a super transmission grid involve citing decisions and providing reasonable policies for equitable cost allocation. In both cases, a congressional response is needed that gives FERC authority to take bold and decisive action (see DOE announcement and http://www.nappartners.com/news/doe-fercannounce-new-collaborative-effort-ontransmission-projects).

10.10   BENEFITS OF SMART TRANSMISSION

Smart transmission investment provides many benefits to power customers and electricity markets. Although they vary according to the type and location of smart technologies, installation of new digital technologies materials and implementation of the functionalities identified in this chapter are aimed at achieving:

- Increased reliability
- Increased electricity throughput at lower delivered cost
- More efficient fuel use for generation, yielding lower air emissions
- Greater use of RER and clean generation resources, with lower operational integration costs.
- More effective use of energy storage, reduce the costs of the peak electricity provision
- Facilitate third party participation in the power system
- Foster wholesale and retail markets, improving information available to customers and market participants about grid connections, electricity prices, and usage

10.11   SUMMARY

This chapter has discussed how the existing transmission system can be made even smarter by investing in new technology. Most of these smart grid technology elements are well-tested, mature, and cost-effective, and their use will make the North American bulk power system more reliable, secure, efficient, economic, diverse, and environmentally sustainable. But while communications, computer analytical tools [8], sensors, and controls are critical elements, such technologies [9] cannot themselves deliver electricity from a power plant to the end-user. The chapter pointed out the need for a strong platform of wires, cables, and substations, investment in existing transmission infrastructure, and additional investment in new wires and transformers. The combination of conventional transmission technologies with advanced smart grid elements that will optimize and enhance the value of transmission investments was explained.

REFERENCES

[1] J.A. Momoh, M.E. El-Hawary, and R. Adapa. "A Review of Selected Optimal Power Flow Literature to 1993, Part I: Nonlinear and Quadratic Programming Approaches," *IEEE Transactions on Power Systems* 1999, 14, 96–104.

[2] J. Si, A.G. Barto, W.B. Powell, and D. Wunsch. *Handbook of Learning and Approximate Dynamic Programming*, Hoboken, NJ: Wiley, 2004.

[3] G.K. Venayagamoorthy, R. G. Harley, and D.C. Wunsch. "Dual Heuristic Programming Excitation Neuro-Control for Generators in a Multi-machine Power System," *IEEE Transaction on Industry Applications* 2003, 39, 382–394.

[4] J.A. Momoh. *Electric Power System Application of Optimization*, New York: Marcel Dekker, 2001.

[5] N.P. Phady. "Unit Commitment—A Bibliographical Survey," *IEEE Transactions on Power Systems* 2004, 19, 1196–1205.

[6] J.A. Momoh and Y. Zhang. "Unit Commitment Using Adaptive Dynamic Programming," *Proceedings of the IEEE Intelligent Systems Application to Power Systems (ISAP) Conference*, 2005.

[7] J.A. Momoh and A.C. Caven. "Distribution System Reconfiguration Scheme Using Integer Interior Point Programming Technique," *Transmission and Distribution Conference and Exposition*, 2003, PES1, 7–12.

[8] Smart Grid Interoperability Standards Project, National Institute of Standards and Technology. January 2010.

[9] L. Karisny. "Will Security Start or Stop the Smart Grid?" *Digital Communities*, November 18, 2010.

11

EPILOGUE

The smart grid initiative, also known as "intelligent grid" or "smart transmission," is presently being pursued throughout the United States by various entities. It is a completely new idea with initial concepts and work undertaken by utilities, academics, and other stakeholders. The development of supporting technologies is being provided mainly through stimulus funding programs administered by the DOE.

Stakeholders understand that the technological development of new components, such as smart meters, geolocational monitoring equipment, new DR including sustainable energy resources, and energy storage mediums must be complemented by developing the requisite communication infrastructure, retrofitting/updating current grid components, and establishing standards for interoperability and integration. Testbed and experimentation environments are presently being undertaken. Stakeholders in the smart grid have also asserted the importance of training the existing workforce and educating future personnel.

As with any cutting-edge technology, many challenges must be met. The list includes:

1. The integration of multiagent system frameworks and adaptive system thinking to improve system operability for congestion management.

Smart Grid: Fundamentals of Design and Analysis, First Edition. James Momoh.
© 2012 Institute of Electrical and Electronics Engineers. Published 2012 by John Wiley & Sons, Inc.

2. The UC problem with stochasticity for network uncertainty challenges using a DSOPF technology. This technique will ultimately provide the robust co-optimization need for planning and operation or resources.

3. Microgrid development with conversion and protection schemes that will facilitate efficiency. This may involve the use of micro electro-mechanical systems (MEMS) and smart protection schemes.

4. Interdisciplinary approaches to development of large-scale smart buildings and components involving electrical engineering, mechanical engineering, material sciences and information systems.

5. Future aerospace automation technology advancements that will be integrated for improvement of quality of technology and design.

6. Advancement of materials and components such as nanotechnology to be utilized for storage technologies.

7. Pricing studies based on dynamics and stochastic models with both technical and human behavior and serious study to justify the marginal cost benefit associated with investment.

8. Standardization of tools, software, and hardware for design and deployment to ensure smart grid acceptability by the public and stakeholders.

9. Training/retraining and education of present and future workforces through the facilitation of cross-discipline academic research supported by utilities.

Overall examples are needed to represent the designs and prototypes of performance metrics that accurately define the new smart grid technology with the attributes discussed in this book as well as those referenced herein through:

1. Increased use of digital information and controls technology to improve reliability, security, and efficiency of the electric grid.

2. Dynamic optimization of grid operations and resources with full cyber security.

3. Deployment and integration of distributed resources and generation, including RER.

4. Development and incorporation of DR, demand-side resources, and energy efficiency resources.

5. Deployment of smart technologies (real-time, automated, interactive technologies that optimize the physical operation of appliances and consumer devices) for metering, communications concerning grid operations and status, and distribution automation.

6. Integration of smart appliances and consumer devices.

7. Deployment and integration of advanced electricity storage and peak-shaving technologies, including PHEVs and thermal-storage air conditioning.

8. Provision to consumers of timely information and control options.

9. Development of standards for communication and interoperability of appliances and equipment connected to the grid, including the infrastructure serving the grid.
10. Identification and lowering of unreasonable or unnecessary barriers to adoption of smart grid technologies, practices, and services.

Once the above is achieved we will be able to boast of a truly intelligent grid with system applications in space, naval ship systems, and smart technology for public housing and energy consuming facilities.

INDEX

Note: Page numbers in *italics* figures; tables are noted with *t*.

Smart Grid: Fundamentals of Design and Analysis, First Edition. James Momoh.
© 2012 Institute of Electrical and Electronics Engineers. Published 2012 by John Wiley & Sons, Inc.

IEEE Press Series on Power Engineering

Printed and bound by CPI Group (UK) Ltd, Croydon, CR0 4YY

17/04/2025

14658910-0001